Microorganisms for Sustainability

Volume 3

Series editor

Naveen Kumar Arora, Environmental Microbiology, School for Environmental Science, Babasaheb Bhimrao Ambedkar University, Lucknow, Uttar Pradesh, India

More information about this series at http://www.springer.com/series/14379

Tapan Kumar Adhya • Banwari Lal • Balaram Mohapatra • Dhiraj Paul • Subhasis Das
Editors

Advances in Soil Microbiology: Recent Trends and Future Prospects

Volume 1: Soil-Microbe Interaction

Springer

Editors
Tapan Kumar Adhya
KIIT School of Biotechnology
Bhubaneswar, Odisha, India

Balaram Mohapatra
Environmental Microbiology Laboratory
Indian Institute of Technology, Kharagpur
West Bengal, India

Subhasis Das
Environmental & Industrial Biotechnology
The Energy & Resources Institute (TERI)
New Delhi, Delhi, India

Banwari Lal
Environmental & Industrial Biotechnology
The Energy & Resources Institute (TERI)
New Delhi, India

Dhiraj Paul
Microbial Ecology Laboratory
National Centre for Cell Science, Pune
Maharashtra, India

ISSN 2512-1901 ISSN 2512-1898 (electronic)
Microorganisms for Sustainability
ISBN 978-981-10-6177-6 ISBN 978-981-10-6178-3 (eBook)
https://doi.org/10.1007/978-981-10-6178-3

Library of Congress Control Number: 2017964109

Printed on acid-free paper

This Springer imprint is published by the registered company Springer Nature Singapore Pte Ltd. part of Springer Nature.
The registered company address is: 152 Beach Road, #21-01/04 Gateway East, Singapore 189721, Singapore

Contents

About the Editors

Dr. Tapan Kumar Adhya is currently the Director, South Asia Nitrogen Centre, New Delhi, and is also working as Professor in the School of Biotechnology, KIIT University, Bhubaneswar, Odisha, India. He has more than 150 publications and 30 book chapters.

Dr. Banwari Lal is currently the Environmental and Industrial Biotechnology Division, The Energy and Resources Institute, New Delhi. He has several publications and patents in the field of hydrocarbon bioremediation technology.

Mr. Balaram Mohapatra is currently doing PhD at IIT Kharagpur.

Dr. Dhiraj Paul is presently working as young scientist in National Centre for Cell Science, Pune, India.

Mr. Subhasis Das is presently working in Environmental and Industrial Biotechnology Division TERI – The Energy and Resources Institute, New Delhi, India.

About the Series Editor

Dr. Naveen Kumar Arora, Phd Microbiology, Professor in Department of Environmental Science, Babasaheb Bhimrao Ambedkar University (a central university), Lucknow, Uttar Pradesh, India, is a renowned researcher in the field of Environmental Microbiology and Biotechnology. His specific area of research is rhizosphere biology and PGPRs. He has more than 50 research papers published in premium international journals and several articles published in magazines and dailies. He is editor of 10 books, published by Springer. He is member of several national and international societies, Secretary General of Society for Environmental Sustainability, a member of editorial board of four journals, and reviewer of several international journals. He is also the Editor-in-Chief of the journal *Environmental Sustainability* published by Springer Nature. He has delivered lectures in conferences and seminars around the globe. He has a long-standing interest in teaching at the PG level and is involved in taking courses in bacteriology, microbial physiology, environmental microbiology, agriculture microbiology, and industrial microbiology. He has been advisor to 118 postgraduate and 8 doctoral students. Recently, he was awarded for excellence in research by the Honorable Governor of Uttar Pradesh. Although an academician and researcher by profession, he has a huge obsession for the wildlife and its conservation and has authored a book, *Splendid Wilds*. He is President of Society for Conservation of Wildlife and has a dedicated website www.naveenarora.co.in for the cause of wildlife and environment conservation.

Soil Microbiology Research in the Coming Decades: Translational Research Opportunities

1

Tapan Kumar Adhya and K. Annapurna

Abstract

Soil microbiology deals with the diverse group of living organisms that resides in the soil. The microflora and fauna function to maintain the soil process and other ecosystem services to keep the soil healthy and functional, and thereby support plant growth. Agriculture and land-use changes including rapid urbanization irrevocably alter the soil microflora – both structurally and functionally. While plant supports microbes in their root region by transferring part of the photosynthate, microbes dwelling in the rhizosphere form the second genome for the plant and provide physiological and ecological fitness to the plant. For meeting the demand for the burgeoning population of the earth, microorganisms provide a green alternative to grow plants, especially crop plants, in a sustainable manner.

Keywords

Soil microbiology · Microbial diversity · Plant microbiome · Rhizosphere · Sustainable agriculture

The *Glossary of Soil Science Terms* (SSSA 2008) defined soil microbiology as the branch of soil science focussing on the diverse group of soil inhabiting microbes with their functions and interaction activities, which influences other living organisms in various ways. Soil, the living epidermis of the planet with its diverse resources, will continue to determine the human security. Intimate interaction of plant, animal and microbial life, inhabiting the soil matrix, drives redox reactions

T. K. Adhya (✉)
KIIT School of Biotechnology, Bhubaneswar, Odisha, India
e-mail: adhyas@yahoo.com

K. Annapurna
Division of Microbiology, ICAR-Indian Agricultural Research Institute, New Delhi, India

T. K. Adhya et al. (eds.), *Advances in Soil Microbiology: Recent Trends and Future Prospects*, Microorganisms for Sustainability 3,
https://doi.org/10.1007/978-981-10-6178-3_1

that regulate biogeochemical cycles of many elements and creates a pool of organic C that very much exceeds the C in the global atmosphere and biosphere. Microbial communities that mediate these redox reactions are now supposed to represent much of the Earth's total biodiversity. However, the structure, function and economic potentials of this soil biosphere component are only beginning to be explored.

Soils, especially along the great rivers' floodplains being the starting point of human civilization, have been exploited since human community turned from gatherer-hunter to organized population depending upon agriculture as the vocation. India being one such country with civilization records dating back to the Indus valley, such soils have been exploited to a great extent. Although, even after 5000 years, these soils are still considered the grain bowl of the Indian subcontinent, over-exploitation has resulted into qualitative decline in the soil quality and health and resulted in a perceived decline in food production. There is an emerging understanding of the importance of microbial communities for soil health with obvious possibilities of reversing the perceived decline in soil quality through microbiological interventions.

The microbial communities or microbiomes of various environments have been explored to understand their ecological role. The plant microbiome is a major determinant of plant health and productivity and has received considerable attention recently (Bulgarelli et al. 2013). Soil-inhabiting plant roots are in direct connection with microbially the most diverse biome on the planet, with estimates of very high bacterial diversity per gram of soil ($\sim10^{14}$–10^{16}). Due to the importance of the soil as the plant habitat, majority of research focuses on the rhizosphere microbiology, even though microorganisms are also able to readily colonize other plant parts. All these microenvironments provide specific biotic and abiotic setting for microorganisms having a correspondingly specific function for the host. Manipulation of the plant microbiome has the potential to reduce plant diseases especially the soilborne ones, increase agricultural production, reduce chemical inputs like fertilizers and pesticides and reduce emissions of greenhouse gases, resulting in more sustainable agricultural practices. Many of the plant-associated microbes are also crucial players in the global biogeochemical cycles. These microbial communities are influenced by geographical location, soil resource, host genotype and cultivation practices. Dynamics of the colonization pattern for the root-associated microbiome across the three niches, viz. endorhizosphere, ectorhizosphere (root surface) and microbiome, in the close vicinity of the roots provide evidence for rapid enrolment of root-associated microbiomes from soil and support a multistep model wherein each root component plays a selective role in the microbiome congregation (Edwards et al. 2015).

Sustainable crop production requires both technical and organizational advances. While pathogens, pests and weeds cause large quantum of pre- and postharvest losses, beneficial symbionts provide the opportunity to improve yield stability, quantity and quality. Research and development for this area of agricultural research is, therefore, both warranted and urgent. Further, major long-term investments in foundational and translational agricultural research are necessary. Soil microorganisms are key players in agroecosystem functioning since they drive primary and secondary production and nutrient cycles (Nannipieri et al. 2002;

Bardgett et al. 2005). N is the most important nutrient limiting crop growth and also the major element supplied by fertilization (Smil 1997). It has been estimated that in several of the intensively cropped systems all over the world, N input exceeds N output by crop uptake by 10–240 kg N ha^{-1} per year (Sutton et al. 2011). It has been calculated that average N fertilization efficiency of crop production rarely exceeds 30% implying high economic, environmental and societal costs (Tilman et al. 2002; Abrol and Adhya 2017). Excess N input changes rates of N transformation, increasing the pathways for N loss; about 50–60% of the N surplus is lost as molecular N (N_2) emission by nitrification-denitrification, followed by ammonia (NH_3) volatilization, nitrate (NO^{3-}) leaching and run-off and nitrous oxide (N_2O) and mononitrogen oxide (NOx) emissions (De Vries et al. 2011; Velthof et al. 2011). Apart from leakage to the environment leading to eutrophication, N fertilizer surplus increases the decomposition rate of crop residues and soil organic matter (SOM), with a net decline in soil carbon (C) content and soil fertility affecting not only two major nutrient cycles but also ecosystem services for ecological sustainability.

Soil microbiomes are complicated, highly diverse ecosystems containing large populations of interacting microorganisms. High-quality reads clustered, using >97% sequence identity of root-associated microbial communities of rice plants sampled at 42 days, grouped into 101,112 microbial OTUs with high abundance of methanogenic archaea (Edwards et al. 2015). Similarly, analysis of the microbiome of disease-suppressive soils categorized over 33,000 bacterial and archaeal OTUs in the sugar beet rhizosphere (Mendes et al. 2011). Recently, development of several powerful metagenomic and bioinformatic analysis techniques has enabled the microbiologists to rapidly sequence and identify DNA extracted from soil samples (Biswas and Sarkar 2017; Lagos et al. 2015; Knief 2014). This has enabled examining the genetics of whole microbial communities in order to probe the physiological characteristics and potential of plant-associated microorganisms (Sessitsch et al. 2012). Amplicon sequence analyses of marker genes, typically 16S rRNA in case of bacteria, enable us to characterize the relative abundance of different species in diverse plant compartments including phyllosphere and rhizosphere (Lundberg et al. 2012). Metatranscriptomic approaches, on the other hand, may be used to examine the metabolic activities and regulatory mechanisms that function in discrete environments (Chaparro et al. 2014; Newman et al. 2016).

Beneficial microbes have long been used in agriculture. For decades farmers have been adding nitrogen-fixing bacteria for growing legume crops and mycorrhizal fungi that help plants acquire nutrients for decades. These groups of microorganisms, which have been broadly termed as biofertilizer, keep the soil environment rich in diverse micro- and macronutrients through nitrogen fixation, phosphate and potassium solubilization or mineralization and their acquisition, production and release of plant growth-regulating substances, production of antibiotics and biodegradation of organic matter in the soil. When biofertilizers are applied as seed or soil inoculants, they multiply and participate in nutrient cycling and benefit crop productivity (Singh et al. 2011). Microbial inoculants have overriding significance in integrated nutrient management systems for sustained agricultural productivity and healthy environment (Adesemoye et al. 2009). Efficient strains of *Azotobacter*, *Azospirillum*, *Phosphobacter* and *Rhizobacter* can provide significant amount of available nitrogen through nitrogen cycling and even enhance fertilizer use efficiency by reducing fertilizer application. The biofertilizers produce several plant hormones including indole acetic acid (IAA), gibberellins (GA) and cytokinins (CK). Biofertilizers are also known to improve photosynthesis to confer plant tolerance to stress and increased resistance to pathogens, thereby resulting in crop productivity improvement and important tool for climate-resilient agriculture (Sahoo et al. 2013; Kashyap et al. 2017).

As mentioned in the beginning, plants by virtue of their growing in soil had access to the most diverse microbiome in the world. When the seed germinates on the soil, it sends its roots into the soil to procure water and mineral nutrients. In this process, the roots engineer soil physical structure and, by releasing root exudates in and around root zone, recruit soil microorganisms as their closest neighbour. At the same time, sloughed off dead and decaying root tissues, processed by the heterotrophic microorganisms, lead to the production of soil organic matter. Thus, there is compelling evidence that plants engineer the rhizosphere microbiome (Chaparro et al. 2014). This is further strengthened by the fact that even the most ancient plant lineages demonstrate a strong ability to alter the relative abundance of microbial groups in the soils surrounding the rhizosphere (Valverde et al. 2016). Due to such intricate relationship whereby plant species support unique microbiomes, possibility arises whether we can manipulate such relationship that has now being named as 'rhizosphere engineering' leading to increased productivity (Akhami et al. 2017). Presently, our ability to manage and manipulate the rhizosphere microbiome is limited to alter the microbiome through inoculation and addition of organic matter to increase the diversity. On several occasions, such introduced microorganism or 'inoculants', as they are popularly termed, fail due to extraneous reasons like predation or getting outcompeted by the microorganisms already selected over a long time period. Possibly, new gene editing and synthetic biology tools offer alternate path to engineer microbes with targeted functions (Wallenstein 2017; Hutchison et al. 2016) or engineering plant traits (Nogales et al. 2015).

Evidence is accumulating that the immense diversity of microorganisms and animals that lie belowground contributes significantly in shaping the aboveground biodiversity and the functioning of terrestrial ecosystems (Bardgett and van der Putten 2014). Globally, understanding of how this belowground biodiversity is distributed and how it regulates the structure and functioning of terrestrial ecosystems is rapidly growing (Carey 2016, albeit at a much limited scale for the tropical ecosystem. Evidence also points to soil biodiversity playing a key role in determining the ecological and evolutionary responses of terrestrial ecosystems to current and future environmental change. Thus, three major areas that would require focussed attention of the scientists during the next decades are:

1. Intensive study on the impact of huge domestication of soils on the total microbial diversity and use of the reverse engineering to restore soil health and fertility. It is now becoming clear that crop management and other anthropogenic interventions in agriculture affects both the structural and functional diversity of the soils.

2. Thorough research on plant microbiome as soil microorganisms found in the root zone impact plant growth and development. However, as the potential to harness these benefits is hampered by sheer abundance and diversity of the players influencing desirable plant traits, focused attention on the study of rhizosphere biology, impacting factors and the complex interaction on the well-being of plant and maintenance of soil health is essential.
3. Organic C, the key driver of soil sustainability, stored in soil is the balance between plant inputs and microbially mediated metabolic losses of CO_2. Modern agriculture is considered a major disruption to the natural C balance in soil, effectively channelizing the microbial mediated processes to release a vast store of labile C that has accumulated over millennia.
4. Current understanding on the overarching impacts of climate change on life processes in the earth including the projected adverse impacts on field agriculture and crop plants is also considered to be influenced by soil microorganisms. Land-use change through cultivation and clearing has caused a major fraction of total anthropogenic greenhouse gas emissions since the nineteenth century. Fortunately, mitigation of such climate processes also appears to be mediated by soil microbes and needs to be intensely investigated and implemented.

The new age ecological understanding that a plant is not an isolated individual at its genomic level but a larger genomic pool inclusive of its associated microbial genome, 'the microbiome', has given rise to the 'holobiont' concept (Guerrero et al. 2013). Similar to ecological systems of higher organisms, the holobiont shows interdependent and complex dynamics (Vandenkoornhuyse et al. 2015). While plants develop from seeds, the microbiome has a multitude of sources majority being recruited from the soil (Pieterse et al. 2016). The assemblage of these communities depends on the interaction between the emerging seedling and its surrounding environment. These microbial communities are controlled by the plant through diverse strategies, such as the specific profile of root exudates and its immune system led by jasmonic acid or otherwise. Despite this control, the microbiome is still able to adapt and thrive. The molecular knowledge behind these interactions and microbial '-omic' technologies are going forward to the point of enabling holobiont engineering. The collective genome of the rhizosphere microbiome is much larger than that of the plant and is referred to as the plant's second genome or pan-genome. It is the holobiont which responds to the various biotic and abiotic stresses rather than the plant alone, as perceived earlier. Hence, the overall fitness of the plant is regulated by the plant itself and its 'microbiome'. This includes the genomic contribution made by the diverse microbial communities that inhabit the surface and internal tissues of the plant parts. New molecular methods have revealed microbiomes to be key components of plant health (Pineda et al. 2017).

Sufficient evidence has now accrued to show that the microbiome mediates several critical plant functional traits, has great significance on plant phenome plasticity and can become a new trajectory for plant neo-domestication. This bears importance in the context of plant breeding strategies as the microbiome offers genetic variability to plants (Edwards et al. 2015). Thus, research needs to focus to co-propagate the co-evolved, i.e. the plant genome and its microbiome.

1. One approach could be the transfer of microbiome from one plant species to other. The overlapping core microbiome between plants gives hope for cross-compatibility of microbiome transfer with phylogenetically unrelated plant species.
2. Development of synthetic communities/microbiome consisting of key players.
3. Transfer of micro-RNA from rhizospheres of donor soils to recipient soils.

Although the plant microbiome is recognized as an important resource pool of microbial diversity, numerous important plant species and their natural relatives have not yet been studied for their associated microbial communities. With an approximate number of 5×10^6 plant species, a considerable amount of work lay ahead of plant microbiome research to explore newer aspects of phylogenetic diversity of plant-associated microorganisms in the future. This might be particularly interesting with plants from extreme environments including aspects of land-use change. Land plants continuously contact beneficial, commensal, and pathogenic microbes in soil via their roots. There is limited knowledge as to how the totality of root-associated microbes (i.e. the microbiome) is shaped by various factors or its pattern of acquisition in the root. The study of plant-microbe associations by new techniques has significantly improved our understanding of the structure and specificity of the plant microbiome. Yet, microbiome function and the importance of the plant's microbiome in the context of human and plant health remain largely unexplored (Mendes et al. 2013).

References

Abrol YP, Adhya TK (2017) Technical summary. In: Abrol YP, Adhya TK, Aneja VP, Raghuram N, Pathak H, Kulshrestha U, Sharma C, Singh B (eds) The Indian nitrogen assessment. Elsevier, Colombo

Adesemoye AO, Torbert HA, Kloepper JW (2009) Plant growth-promoting rhizobacteria allow reduced application rates of chemical fertilizers. Microb Ecol 2009(58):921–929

Ahkami AH, White RA III, Handakumburaa PP, Jansson C (2017) Rhizosphere engineering: enhancing sustainable plant ecosystem productivity. Rhizosphere 3(2017):233–243

Bardgett RD, van der Putten WH (2014) Belowground biodiversity and ecosystem functioning. Nature 515:505–511

Bardgett RD, Usher MB, Hopkins DW (2005) Biological diversity and function in soils. Cambridge University Press, Cambridge

Biswas R, Sarkar A (2017) "Omics" tools in soil microbiology: the state of the art. In: Adhya TK, Lal B, Mahapatra B, Paul D, Das S (eds) Advances in soil microbiology: recent trends and future prospects: Vol. 1: Soil-microbe interaction. Springer, New Delhi, pp 35–64

Bulgarelli D et al (2013) Structure and functions of the bacterial microbiota of plants. Annu Rev Plant Biol 64:807–838

Carey J (2016) Crucial role of belowground biodiversity. PNAS 113:7682–7685

Chaparro JM, Badri DV, Vivanco JM (2014) Rhizosphere microbiome assemblage is affected by plant development. ISME J 8:790–803 de Vries W, Leip A, Reinds GJ, Kros J, Lesschen JP, Bouwman AF (2011) Comparison of land nitrogen budgets for European agriculture by various modeling approaches. Environ Pollut 159:3254–3268

Edwards J, Johnson C, Santos-Medellín C, Lurie E, Podishetty NK, Bhatnagar S, Eisen JA, Sundaresan V (2015) Structure, variation, and assembly of the root-associated microbiomes of rice. PNAS 112:E911–E920

Guerrero R, Margulis L, Berlanga M (2013) Symbiogenesis: the holobiont as a unit of evolution. Int Microbiol 16:133–143

Hutchison CA, Chuang RY, Noskov VN, Assad-Garcia N, Deerinck TJ, Ellisman MH et al (2016) Design and synthesis of a minimal bacterial genome. Science 351. https://doi.org/10.1126/science.aad6253

Kashyap PL, Rai P, Srivastava AK, Kumar S (2017) Trichoderma for climate resilient agriculture. World J Microbiol Biotechnol 33:155. https://doi.org/10.1007/s11274-017-2319-1

Knief C (2014) Analysis of plant microbe interactions in the era of next generation sequencing technologies. Front Plant Sci 5:Article 216. https://doi.org/10.3389/fpls.2014.00216

Lagos ML, Maruyama F, Nannipieri P, Mora ML, Ogram A, Jorquera MA (2015) Current overview on the study of bacteria in the rhizosphere by modern molecular techniques: a mini–review. J Soil Sci Plant Nutr 15(2):504–523

Lundberg DS, Lebeis SL, Paredes SH, Yourstone S, Gehring J, Malfatti S, Tremblay J, Engelbrektson A, Kunin V, del Rio TG, Edgar RC, Eickhorst T, Ley RE, Hugenholtz P, Tringe SG, Dang JL (2012) Defining the core Arabidopsis thaliana root microbiome. Nature 488:86–90

Mendes R, Kruijt M, de Bruijn I, Dekkers E, van der Voort M, Schneider JHM, Piceno YM, DeSantis TZ, Andersen GL, Bakker PAHM, Raaijmakers JM (2011) Deciphering the rhizosphere microbiome for disease-suppressive bacteria. Science 332:1097–1100

Mendes R, Garbeva P, Raaijmakers JM (2013) The rhizosphere microbiome: significance of plant beneficial, plant pathogenic, and human pathogenic microorganisms. FEMS Microbiol Rev 37:634–663

Nannipieri P, Ascher J, Ceccherini M, Landi L, Pietramellara G, Renella G (2002) Microbial diversity and soil. Eur J Soil Sci 54:655–670

Newman MM, Lorenz N, Hoilett N, Lee NR, Dick RP, Liles MR, Ramsier C, Kloepper JW (2016) Changes in rhizosphere bacterial gene expression following glyphosate treatment. Sci Total Environ 553:32–41

Nogales A, Valadas V, Ragonezi C, Polidoros A, Arnholdt-schmitt B (2015) Can functional hologenomics aid tackling current challenges in plant breeding? Brief Funct Genomics:1–10. https://doi.org/10.1093/bfgp/elv030

Pieterse CMJ et al (2016) The soil-borne supremacy. Trends Plant Sci 21:171–173

Pineda A, Kaplan I, Bezemer T (2017) Steering soil microbiomes to suppress aboveground insect pests. Trends Plant Sci 22:770–778

Sahoo RK, Bhardwaj D, Tuteja N (2013) Biofertilizers: a sustainable eco-friendly agricultural approach to crop improvement. In: Tuteja N, Gill SS (eds) Plant acclimation to environmental stress. Springer, New York, pp 403–432

Sessitsch A, Hardoim P, Döring J, Weilharter A, Krause A, Woyke T, Mitter B, Hauberg-Lotte L, Friedrich F, Rahalkar M, Hurek T, Sarkar A, Bodrossy L, van Overbeek L, Brar D, van Elsas JD, Reinhold-Hurek B (2012) Functional characteristics of an endophyte community colonizing rice roots as revealed by metagenomic analysis. Mol Plant-Microbe Interact 25:28–36

Singh JS, Pandey VC, Singh DP (2011) Efficient soil microorganisms: a new dimension for sustainable agriculture and environmental development. Agric Ecosyst Environ 140:339–353

Smil V (1997) Global population and the nitrogen cycle. Sci Am 277:76–81

SSSA [Soil Science Society of America] (2008) Glossary of soil science terms. ASA-CSSA-SSSA, Madison

Sutton MA, Howard CM, Erisman JW, Billen G, Bleeker A, Greenfelt P et al (eds) (2011) The European nitrogen assessment. Cambridge University Press, Cambridge

Tilman D, Cassman KG, Matson PA, Naylor R, Polasky S (2002) Agricultural sustainability and intensive production practices. Nature 418:671–677

Valverde A, De Maayer P, Oberholster T, Henschel J, Louw MK, Cowan D (2016) Specific microbial communities associate with the rhizosphere of *Welwitschia mirabilis*, a living fossil. PLoS One 11:e0153353

Vandenkoornhuyse P, Quaiser A, Duhamel M, Le Van A, Dufresne A (2015) The importance of the microbiome of the plant holobiont. New Phytol 206:1196–1206

Velthof G, Barot S, Bloem J, Butterbach-Bahl K, de Vries W, Kros J, Lavelle P, Olesen EJ, Oenema O (2011) Nitrogen as a threat to European soil quality. In: Sutton MA, Howard CM, Erisman JW, Billen G, Bleeker A, Grennfelt P, van Grinsven H, Grizzetti B (eds) The European nitrogen assessment. Cambridge University Press, Cambridge, pp 495–510

Wallenstein MD (2017) Managing and manipulating the rhizosphere microbiome for plant health: a systems approach. Rhizosphere 3:230–232

2 Molecular Genomic Techniques for Identification of Soil Microbial Community Structure and Dynamics

Dhiraj Paul, Satish Kumar, Mrinal Mishra, Sushant Parab, Sunil Banskar, and Yogesh S. Shouche

Abstract

Soil microorganisms play a crucial role in maintaining major biogeochemical/nutrient cycle, soil quality, and productivity. Hence, the understanding of soil microbial community structure, distribution, and their metabolic function is essential for getting a deeper insight into soil ecosystem and its health. A number of molecular methods for extracting metagenome, total RNA, protein, and metabolites from the diverse environmental samples, sequencing technology, etc. are present which help to know about microbial structure, composition, and their metabolic function in the specific environmental ecosystem. Genetic fingerprinting like ARDRA, RFLP, DGGE, and T-RFLP and omics approaches like metagenomics, metatranscriptomics, and metabolomics are essential techniques for identifying and depicting the total microbial community structure and their interactions with environmental and biotic factors. So for these molecular techniques, it is possible to identify and functionally characterize soil microbes that are not culturable in a laboratory environment. This chapter describes old and modern novel state of the art molecular techniques which proved insights into the phylogenetic and functional activities of microbial assemblages in a terrestrial ecosystem.

D. Paul (✉) · M. Mishra · S. Parab · S. Banskar · Y. S. Shouche (✉)
Microbial Culture Collection, National Centre for Cell Science, Savitribai Phule University of Pune Campus, Ganeshkhind, Pune, Maharashtra, India
e-mail: pauldhiraj09@gmail.com; yogesh@nccs.res.in

S. Kumar
Microbial Culture Collection, National Centre for Cell Science, Savitribai Phule University of Pune Campus, Ganeshkhind, Pune, Maharashtra, India

ICAR-National Institute of Abiotic Stress Management, Baramati Pune, Maharashtra, India

T. K. Adhya et al. (eds.), *Advances in Soil Microbiology: Recent Trends and Future Prospects*, Microorganisms for Sustainability 3,
https://doi.org/10.1007/978-981-10-6178-3_2

Keywords
Microbial ecology · Stable isotope probing · Autoradiography · Fluorescence *in situ* hybridisation · DGGE · Next generation sequencing

2.1 Introduction

In soil/natural ecosystems, microorganisms including bacteria and fungi exist in a very large number and play a very crucial role in maintaining major biogeochemical cycles (Molin and Molin 1997; Wall and Virginia 1999), plant nutrition (George et al. 1995; Timonen et al. 1996), plant health (Srivastava et al. 1996; Filion et al. 1999; Smith and Goodman 1999; Wright and Upadhyaya 1998; Dodd et al. 2000), soil fertility (Yao et al. 2000; O'Donnell et al. 2007), soil structure (Wright and Upadhyaya 1998), and degrading organic pollutants and remediation of toxic metals (Barakat 2011). Therefore, microorganisms are key players in important ecological processes, such as carbon, nitrogen, phosphorous, and sulfur biogeochemical cycle, and directly influenced all lives on Earth (Garbeva et al. 2004). It is noted that 1 gm of soil/sediment may contain 10^9 bacterial cell (Whitman et al. 1998). In terrestrial environments, soil sustains as many as 4–5 $\times$ 10^{30} microbial cells and in aquatic environments approximately 1.2 $\times$ 10^{29}cell (Whitman et al. 1998; Singh et al. 2009). It constitutes 60% of the total biomass of the Earth, and it represents two to three orders greater biomass than the total plant and animal cells (Singh et al. 2009). Therefore, a large number of microorganisms and their genetic diversity are unexplored, and that is directly involved in maintaining major nutrient cycles, global climate change, and the greenhouse effect. So understanding this unexplored genetic diversity is a high-priority issue in microbial ecology.

The soil microbial ecology analysis does not only mean the identification of total microbial biomass and community diversity, but it also explores microbial growth, function, distribution, and interactions among species. Therefore, soil microbial ecologist tries to answer fundamental questions, i.e., (1) What is microbial community structure and composition? (2) What are the metabolic functions/functional genes expressed so that microorganisms can run major biogeochemical cycle in the ecosystem? (3) How do the functional activities of the microorganisms relate to major ecosystem functions including biogeochemical cycling, energy flow, etc.? Besides, anthropogenic activities including city development, agriculture, pesticide use, and other pollution directly affect the soil microbial diversity. How these changes affect surface and subsurface ecosystems is unknown. For a decade many new molecular approaches like NGS, metaproteomics, metabolomics, etc. have evolved that incredibly help soil microbial ecologist for better assessments of microbial diversity and their function in the ecosystem. In-depth understanding of microbial distribution, their function, and interaction helped in the development of new techniques for bioremediation, energy generation processes, pharmaceuticals, food, chemical, mining, etc. Therefore, for addressing how microbial community structure and dynamics affect the ecosystem function, reliable and accurate techniques of soil microbial ecology are needed. In the following sections, we are describing the traditional molecular techniques, current methods, and their advantages and disadvantages which are used for studying soil microbial structure and function.

2.2 Culture Based Techniques: Advantages and Limitation

A diverse group of microorganisms are present in the environment including soil, and for their isolation, identification, characterization, and culture-based microbial diversity analysis purpose, different standard culture-based techniques (Hugenholtz 2002) are available that include use of different types of growth media, namely, Luria-Bertani, nutrient agar, tryptic soy agar, etc., for copiotrophic bacterial growth and R2A, RAVAN, minimal media, and synthetic marine water/groundwater media for oligotrophic bacteria. Despite a number of ways like mimicking the environmental niches (from where the samples are collected) by changing parameters like temperature, pH, nutrient composition, and trace nutrient composition, more than 99% organisms are still uncultivable which are seen as viable under a microscope. More than 20 phyla are present as a candidate division like TM7, OP10, OP11, WS2, WS3, etc. which are taxonomically well defined based on their metagenomics information, but they are still uncultivable (Schloss and Handelsman 2004). Therefore, their ecological and industrial application is not possible due to uncultivable nature. Bacterial phyla *Acidobacteria* which constitute more than 20% soil bacterial population but very few genera of these phyla are culturable, mostly uncultivable. Therefore, for understanding who are present in the ecosystem and what is their role in maintaining major biogeochemical cycle in the particular ecosystem, application of culture-independent molecular techniques is highly desirable.

2.3 Classical Molecular Methods of Microbial Community Analyses

2.3.1 Clone Library Method

Before next-generation/high-throughput sequence-based microbial diversity analysis, a most widely used technique was clone library-based analysis, where PCR product amplified from diverse environment DNA samples is subjected to clone and sequenced the individual clones containing gene fragments. Then for taxonomic assignment, sequences are compared with different databases like Greengene, Ribosomal Database Project (RDP), SILVA, etc. Based on good-quality sequence size, cloned sequences are assigned at a lower taxonomic level like genus and species, but one of the limitations of this technique is being time-consuming and labor intensive. One of the studies showed that environmental samples like soil/sediment may require over 40,000 clones to document 50% of the richness (Dunbar et al. 2002). Generally, using clone library-based approach, one can handle nearly about thousand clones at a time; therefore it gives a tiny picture of the microbial community structure of the particular environment. Many studies are there where this technique is used for identifying the total community structure. For example, recently Pascual and co-worker (2016) used clone library-based techniques for analysis of bacterial communities associated with the rhizosphere of wild plant species found in natural settings where bacterial phyla *Proteobacteria*, *Actinobacteria*, *Acidobacteria*, and *Gemmatimonadetes* dominated. Microbial diversity analysis of hydrocarbon-contaminated sediment samples of northwest of Bemidji, Minnesota, USA, using clone library-based analysis revealed presence of iron-reducing *Betaproteobacteria* followed by *Deltaproteobacteria*, *Smithella*, and the hydrogenotrophic *Methanoregula* (Beaver et al. 2016).

2.3.2 Genetic Fingerprinting Techniques

Difference in genomic or nucleotide sequences can be utilized to generate the identity of any organisms, and the techniques, which facilitate this kind of identification, are known as genetic fingerprinting techniques. Hence, using genetic fingerprinting techniques like ARDRA, DGGE, and T-RFLP, microbial community profiling is done where PCR product amplified from direct environmental DNA samples is used (Table 2.1).

2.3.2.1 Amplified Ribosomal DNA Restriction Analysis (ARDRA)

ARDRA is a method similar to restriction fragment length polymorphism (RFLP) or an extended form of it and originally developed by Vaneechoutte et al. (1993). It was firstly used for characterization of *Mycobacterium* species. Then it was used to characterize other bacterial species also (Vaneechoutte et al. 1995). Previously ARDRA was used for selection of clone libraries and strain typing to determine phylogenetic groups inside a microbial community and to study microbial diversity. This procedure involves amplification of the conserved region of 16S rRNA gene using universal or genus-/species-specific primers through polymerase chain reaction followed by enzymatic digestion of the PCR products. The restricted fragments are segregated on agarose or polyacrylamide gel, and the emerging profile of bands is used for grouping of the community as per genotype or for strain typing (Tiedje et al. 1999). Generally, for 16S rRNA gene product digestion (1.5 kb), tetra cutter restriction enzymes (e.g., MspI, HaeII) are used. Due to random prevalence of the restriction sites, the chance of occurring specific restriction sites of tetra cutter enzyme is 256 bp. Therefore, care should be taken during restriction enzymes selection. The restriction enzymes that possess the same recognition sequence, i.e., isoschizomers, should not be used; otherwise, it will create difficulty in analysis. Although 16S rRNA gene is a promising marker for the differentiation up to species level, ARDRA is very much useful among the groups which have more interspecies similarity to each other (Heyndrickx et al. 1996). ARDRA method is widely used for the discrimination of isolates undergoing different changes and from different environments (Błaszczyk et al. 2011). ARDRA is also helpful in finding out

Table 2.1 Comparison among the genetic fingerprinting techniques

	ARDRA	DGGE	RAPD	T- RFLP	SSCP	RISA
Use of restriction enzyme	Yes	No	No	Yes	No	No
PCR amplification of 16S rRNA gene	Yes	Yes	Yes	Yes	Yes	Amplification of spacer region
GC clamp	Not needed	Needed	Not needed	Not needed	Not needed	Not needed
Advantage	Resolving intraspecies similarity	Less time-consuming	Very quick and easy to assay	Rapid method	Simple and straightforward technique compared to DGGE	Now automated ARISA is used for microbial community analysis
	Microbial composition changes in diff. environment	Very good for comparing the different samples of microbial profile	Random primer used for amplification; therefore no prior knowledge about sequences are needed	Multiple samples are analyzed simultaneously		
Limitation	Laborious, time-consuming	Resolution quality is not good, not able to classify at lower taxonomic level of the organisms	Low reproducibility	If two or more sequences share same terminal restriction site, it shows same peak on electropherogram	Need sequencing data for designing specific primers	Time-consuming and cumbersome
			It required purified DNA for amplification	Many times false terminal restriction fragments generated which result in false peaks being generated	For reproducibility, need highly standardized electrophoretic condition	

structural changes which are undergoing in microbial communities; however, it can't measure microbial diversity or identify specific phylogenetic cluster within a community fingerprinting profile (Liu et al. 1997). Major limitations of this technique are being time-consuming and laborious, and the restriction profile obtained from complex microbial communities is hard to analyze sometimes. Gulitz et al. (2013) compared four water kefirs and found that they consisted of different proportions of genera *Lactobacillus*, *Leuconostoc*, *Acetobacter*, and *Gluconobacter*. Shehata (2012) used this method for characterization of *Lactobacillus* sp. from fermented millet drink and fresh and raw cow milk.

2.3.2.2 Denaturing Gradient Gel Electrophoresis (DGGE)

The technique of DGGE was originally invented by Fischer and Lerman (1980), and for small ribosomal subunit (i.e., 16S rRNA), it was described by Muyzer et al. (1993). For performing DGGE a gradient of chemical denaturant is formed which is in progressively increasing concentration. PCR products have to pass through this gradient in a polyacrylamide gel. On reaching threshold concentration, the PCR products begin to melt and weaker melting domains melt very fast; therefore migration slows intensely. Amplicon which has different sequence composition will migrate differently and stop at various positions in the gradient result in the formation of different band patterns. In DGGE, forward primer is tagged with a GC-rich nucleotide sequence (30–50 bp) as it renders complete separation of ds PCR product into single strand during electrophoresis. For taxonomic identification, bands from gel are excised, reamplified, and sequenced. DGGE has been extensively applied for elucidation of the microbial community structure and finds out changes in microbial community and dynamics of contaminated soil, water, and many other microcosm-based studies (Macnaughton et al. 1999; Ralebitso et al. 2000; Watanabe et al. 2001; Cummings et al. 2003). The key advantage of DGGE is that it allows the observance of the spatial and time-based changes in microbial community structure and gives a clear picture of the prevailing microbial species present in a particular sample (Malik et al. 2008). Major drawbacks of this technique are as follows: (1) several DNA bands/fragments may have the same melting points; (2) the lengths of the DGGE bands are many times smaller in size, that leads to difficulty in proper taxonomic identification; and (3) due to sequence heterogeneity between multiple rRNA operons of one bacterium, it can lead to several bands in DGGE, resulting in overestimation of the diversity. DGGE is a powerful tool which can discriminate among the microbial populations from different ecosystems. DGGE and pyrosequencing can also be used for elucidation of microbial diversity or composition from different environments such as native plants or nursery-raised plants or bulk sediment from mangrove. The results show that DGGE is a vigorous and practical method and effective in discriminating among earlier defined groups (Cleary et al. 2012). Ivone and Conceição (2013) used DGGE and culture-dependent method to identify the bacterial community composition especially from tap water. The results revealed that the members of *Alpha*-, *Beta*-, and *Gammaproteobacteria* were the major lineages.

2.3.2.3 Random Amplified Polymorphic DNA (RAPD)

RAPD is a PCR-based method, where single short oligonucleotide primers (6 bp), i.e., arbitrarily selected, are used for PCR amplification. As primers are short sized, it can anneal arbitrarily at many sites on the genomic/total DNA due to low annealing temperature (~35 °C) (Franklin et al. 1999). Results are obtained from random amplification of various length products from a single reaction. Depending on the microbial community structure/complexity, different band pattern is generated during agarose/polyacrylamide gel electrophoresis. Unlike conventional PCR, RAPD does not require any specific knowledge about targeting organisms. By using a single primer of random nucleotide sequence in a PCR-based method, this process can detect single nucleotide polymorphism. Due to its easy use, it is widely used for genetic fingerprinting of microbial community composition and closely related microbial species and strains. The *Fasciola hepatica* and *F. gigantic* both liver flukes are parasitic trematodes which belong to phylum Platyhelminthes and coexist in part of Africa and Asia. Life cycles of both are similar but possess different transmission characteristic. McGarry and co-workers (2013) have successfully identified these two species using RAPD-based PCR over a period of 12 years from different countries. RAPD was also used as a typing method for *Campylobacter* species from ducks and duck-related environmental samples from Penang, Malaysia (Adzitey et al. 2012).

2.3.2.4 Terminal Restriction Fragment Length Polymorphism (T-RFLP)

This method (T-RFLP) is used for the exploration of complex microbial community based on the occurrence of recognition sequence of restriction enzymes on the 16S rRNA gene. It is one of the fingerprinting methods aimed to understand the unknown microbial community. The method was invented by Liu et al. (1997). This method includes fluorescent labeling of end of PCR products and restriction digestion of the PCR products of different variants of a single gene. For amplification one or both primers should have their 5′ end labeled with a fluorochrome molecule. Fluorescent molecules, namely, TAMARA, HEX, or 6-FAM, can be used for 5′ end labeling. The mixture of amplicon is subjected to restriction digestion by using one or more restriction enzymes (generally tetra cutter). After the restriction digestion, fragments are separated in a DNA sequencer either by capillary or by polyacrylamide electrophoresis, and the fluorescence detector in the DNA sequencer determines the sizes of the different terminal fragments (Osborn et al. 2000). Due to the use of dye, only fluorescently labeled terminal fragments are detected and determined, while all other fragments are not considered. Therefore, T- RFLP method is different from ARDRA or RFLP where all the fragments are visualized. This procedure also includes purification of PCR product before performing restriction digestion or if a capillary electrophoresis is being used, then before running the sample desalting is also done. The obtained results will be found in graph form, called electropherogram, where Y axis denotes the fluorescence intensity of each fragment and X axis denotes the fragment size. Thus, the bands which appear on an electrophoresis gel are visualized as a peak on the electropherogram. In a T-RFLP, each genetic variant in the original sample is supposed to represent as a single peak, whereas peak height and area represent its relative abundance in a particular community.

2.3.2.5 Single-Strand Conformation Polymorphism (SSCP)

It is a simple and sensitive method to detect polymorphism in DNA. It utilizes the variation in single nucleotide sequences of identical length that can arise under certain conditions such as mutation or single nucleotide polymorphism. This allows separation of different fragments due to their different conformation by using gel electrophoresis and ultimately helps in distinguishing different sequences. The technique was first described by Masato Orita et al. (1989). In SSCP, the environmental DNA is first amplified using PCR and then denatured. After denaturation, single-stranded DNA is separated on polyacrylamide gel (Schwieger and Tebbe 1998). Even a minute difference (often a single base pair) can result into different secondary structures, migrating differently in the gel leading to separation of different sequences in the form of different bands. The technique works on the principle that under non-denaturing conditions, DNA can form different secondary structures based on specific sequences. Molecules having a minute difference like single base substitution may generate different conformers and migrate differently in non-denaturing polyacrylamide gel (Sheffield et al. 1993). Goszczynski (2007; Goszczynski and Jooste 2015) used SSCP method to check the heterogeneity of grape wine virus A and found that it is a rapid and relatively low-cost preliminary analysis of molecular heterogeneity of viruses. Generally, bacteria are involved in spoilage of processed food products, but some fungi are also responsible for this. Dorn-in et al. (2013) tested the presence of fungi in heat-processed meat product using PCR-SSCP. The result shows the presence of *Aureobasidium pullulans*, *C. tropicalis*, *C. zeylanoides*, and *Pichia membranifaciens* and/or species such as *Guignardia mangiferae*, *Lewia infectoria*, and *Lasiodiplodia theobromae*. Pure cultures of *Pseudomonas fluorescens*, *Sinorhizobium meliloti*, and *Bacillus subtilis* have also been successfully differentiated using SSCP (Schwieger and Tebbe 1998).

2.3.2.6 Ribosomal Intergenic Spacer Analysis (RISA)

In this method PCR amplification of a region of 16S rRNA gene known as an ISR, i.e., intergenic spacer region, is generally done. Spacer region is found between large 23S and small 16S subunit of rRNA operon. A significant heterogeneity in ISR region in terms of nucleotide and length is noted. RISA fragments can be generated with the help of oligonucleotide primers which are complimentary to 23S and 16S rRNA genes. The resulting PCR products will be a mixture of fragments representing many dominant community members. These fragments are representing most of the dominant bacteria in an environmental sample. RISA is used originally to explore microbial diversity in soils. It is also used for monitoring microbial community composition in anaerobic treatment plants or bioreactors (Ciesielski et al. 2013). Besides, an automated form of RISA, i.e., ARISA, was used for bacterial community composition analysis of freshwater system (Fisher and Triplett 1999). Although RISA is one of the virtuous methods for the analysis of microbial community composition, the limitation of this method is the same as conventional PCR like primer mismatch, annealing timing, DNA concentration and quality, etc. ARISA techniques were used to detect the presence of different types of *Clostridium* species in raw tank milk and curd used for cheese production in dairies situated in different parts of Northern Padan Plain (Feligini et al. 2015).

2.4 Modern Molecular Methods of Microbial Community Analysis

2.4.1 Stable-Isotope Probing (SIP)

Stable-isotope probing (SIP) technique has become the state of the art in microbial ecology for identifying and detecting microorganisms that are actively involved in specific metabolic processes and elemental fluxes taking place in environmental samples in order to effectively link the taxonomic identity with function (Vogt et al. 2016). This is the most widely employed technique to identify and characterize active community members or specific functional groups of microbial communities that are capable of utilizing specific isotopic-labeled substrates. SIP basically tracks the incorporation of heavy stable isotopes by incubating an environmental sample to substrates containing ^{13}C, nitrogen (^{15}N), ^{3}H, and ^{18}O that are assimilated into microbial biomass of environmental samples. The isotopically labeled carbon/ hydrogen/nitrogen from the substrate gets incorporated into the biomass (particularly DNA, RNA, and proteins) of the active microorganisms in the sample and serves as biomarkers of active community members. After stable isotopes have been assimilated in the environmental sample and metabolically active cells, the label goes into their biomass including DNA, RNA, lipid, and proteins. These labeled biomolecules serve as biomarkers, which are recovered and analyzed using various techniques like fingerprinting, microarrays, clone libraries, metagenomics, and next-generation sequencing (Uhlik et al. 2013). Depending on the type of the isotopic label incorporated and biomarker recovered as a target in a particular study, the SIP can be categorized as DNA-SIP, RNA-SIP, and protein-SIP. However, DNA-SIP, using ^{13}C-based isotopic labels, has been the most extensively used strategy to decipher the microbial populations with a defined function in different sorts of environmental samples (Radajewski et al. 2002; Neufeld et al. 2008; Antony et al. 2010; DeRito et al. 2005; Zhang et al. 2016). The advantage of the DNA-SIP-based studies is that the recovered labeled DNA after the SIP experiment can be subjected to several downstream analyses. A great wealth of significant genetic information is provided by the metagenomes of organisms which include characterization of metabolism-related functional genes and also the ribosomal genes using various fingerprinting-based approaches (e.g., denaturing gradient gel electrophoresis (DGGE), clone libraries, metagenomic libraries, and high-throughput sequencing). A typical DNA-SIP experiment begins with incubation of an environmental sample with labeled substrate in microcosm, maintaining the environmental conditions to the mimicking extent. Incubations for DNA-SIP can be performed either in vitro using laboratory microcosms constructed from field-collected samples (Wald et al.

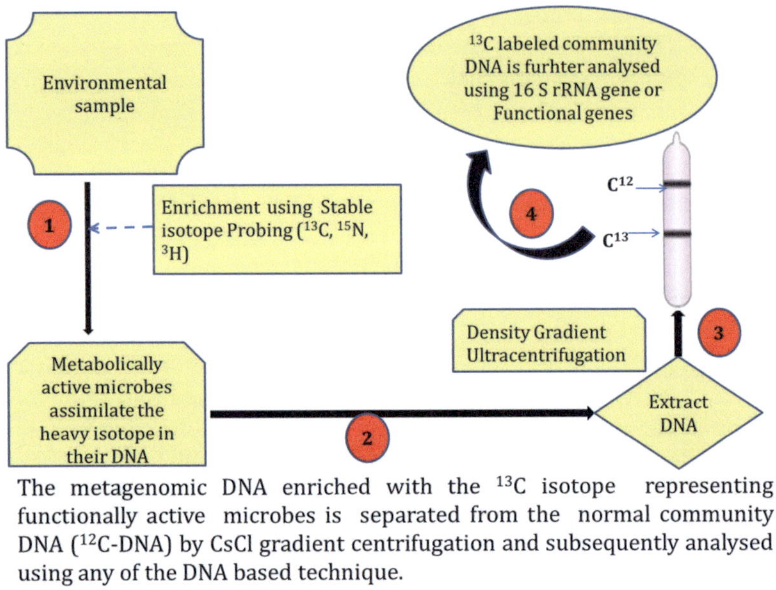

Fig. 2.1 An outline of a typical DNA-SIP experiment

2015; Paes et al. 2015) or in situ using incubation directly in soil or sediment (Liou et al. 2008; Key et al. 2013). Substrate concentration and exposure time always remain to decide critical factors and hence must be for sufficient time so that detectable amount of label gets incorporated in the sample. However, care must be taken to avoid excessive labeling, as the label may rapidly spread, via trophic interactions and cross-feeding (Neufeld et al. 2007). The control samples exposed to the unlabeled substrate should always be included in order to confirm that the DNA recognized as being isotopically labeled is truly the result of labeling with stable isotope and not due to the difference in GC content. The outline flowchart of a representative DNA-SIP experiment is presented in Fig. 2.1; however, the details of all the requirements, steps involved in DNA-SIP protocol, and critical technical considerations can be found in Neufeld et al. (2007) and Dunford and Neufeld (2010).

Several studies suggest the superiority of RNA as biomarker in SIP (RNA-based SIP or RNA-SIP) due to the remarkable attributes of RNA like higher synthesis rate, direct reflection of cellular activity, replication-independent turnover in the cell, and more responsiveness to environmental conditions (Manefield et al. 2002; Whiteley et al. 2007). RNA-SIP is technically more demanding as RNA-SIP requires a more rigorous procedure for gradient evaluation than DNA-SIP. RNA-SIP cannot be performed in cesium chloride (CsCl) gradient media, as CsCl itself precipitates at the buoyant density required for rRNA, and hence, the use of cesium trifluoroacetate (CsTFA) can be a better choice for gradient formation (Rickwood 1992; Manefield et al. 2002). RNA-SIP has been successfully used to decipher the active community members in various types of environmental samples like bioreactor sludge (Manefield et al. 2002), paddy soil (Lueders et al. 2004), rice rhizosphere (Lu and Conrad 2005), and grassland soil (Rangel-Castro et al. 2005).

Nucleotide-based SIP approaches like DNA-SIP and RNA-SIP strictly require an incorporation of around 20% ^{13}C label (Radajewski et al. 2000), while protein-SIP-based approach is 200-fold more superior in sensitivity and can detect incorporation levels of ^{13}C below 1% (Taubert et al. 2011). In a typical protein-SIP-based study, the environmental samples are incubated with stable isotopes (^{13}C, ^{15}N, ^{36}S) which are subsequently incorporated into the amino acids, peptides, and proteins. The amount of atoms replaced by their heavy isotopes changes the natural isotope composition of the labeled peptides. The rate of incorporation of stable isotopes is used for assessing the metabolic activity of the corresponding species (Jehmlich et al. 2008, 2012). The incorporation of a heavy isotope in the proteins is detected by high-resolution mass spectrometry (HRMS) and nano-secondary ionization mass spectrometry (nano-SIMS). The subsequent analysis takes into account the relative isotopic abundance (RIA), and calculation of the RIA is either done by analysis of the distribution of different isotope patterns of the peptides or based on features of the peptide mass such as the relation of the parent mass to the first two digits (Seifert et al. 2012).

2.4.2 FISH (Fluorescence In Situ Hybridization)

FISH is an excellent technique for reliable and rapid identification of microorganisms from environmental samples, and prokaryotic cells can rapidly be identified without cultivation using FISH. FISH involves hybridization of oligodeoxynucleotide complementary to ribosomal RNA sequences (rRNA-targeted nucleic acid probes) that have phylogenetic group-specific sequence signatures. In laboratory protocol of FISH, whole cells from environmental samples to be studied are often fixed by ethanol or paraformaldehyde treatment, and their 16S or 23S rRNA is hybridized with fluorescently labeled taxon-specific oligonucleotide probes. The labeled cells are viewed by scanning confocal laser microscopy (SCLM) (Hill et al. 2000). The abundance of ribosomes (104–105) per cell and consequent abundance of rRNA gene in bacterial cell, apparently observed lack of lateral gene transfers, and a good length of about 1500 and 3000 nucleotides for 16S and 23S, respectively, serve as a basis for hybridization of group-specific fluorescent probes complimentary to rRNA gene. Further, the parameters, such as probe length, GC content, and targeted region of the gene, are the crucial factors for deciding the sequence of the correct species and genus-specific 16S/23S rRNA probes. A workflow of the typical FISH experiment on environmental samples is presented in Fig. 2.2. As in FISH experiment, whole cells are hybridized with group-specific probes, the artifacts and bias introduced due to the DNA extraction, PCR artifacts, and cloning are avoided (Felske et al. 1998). FISH has been successfully applied to study the microbial community composition of different environmental samples (Müller et al. 2016; Kindaichi et al. 2016), and several studies are reported in soil

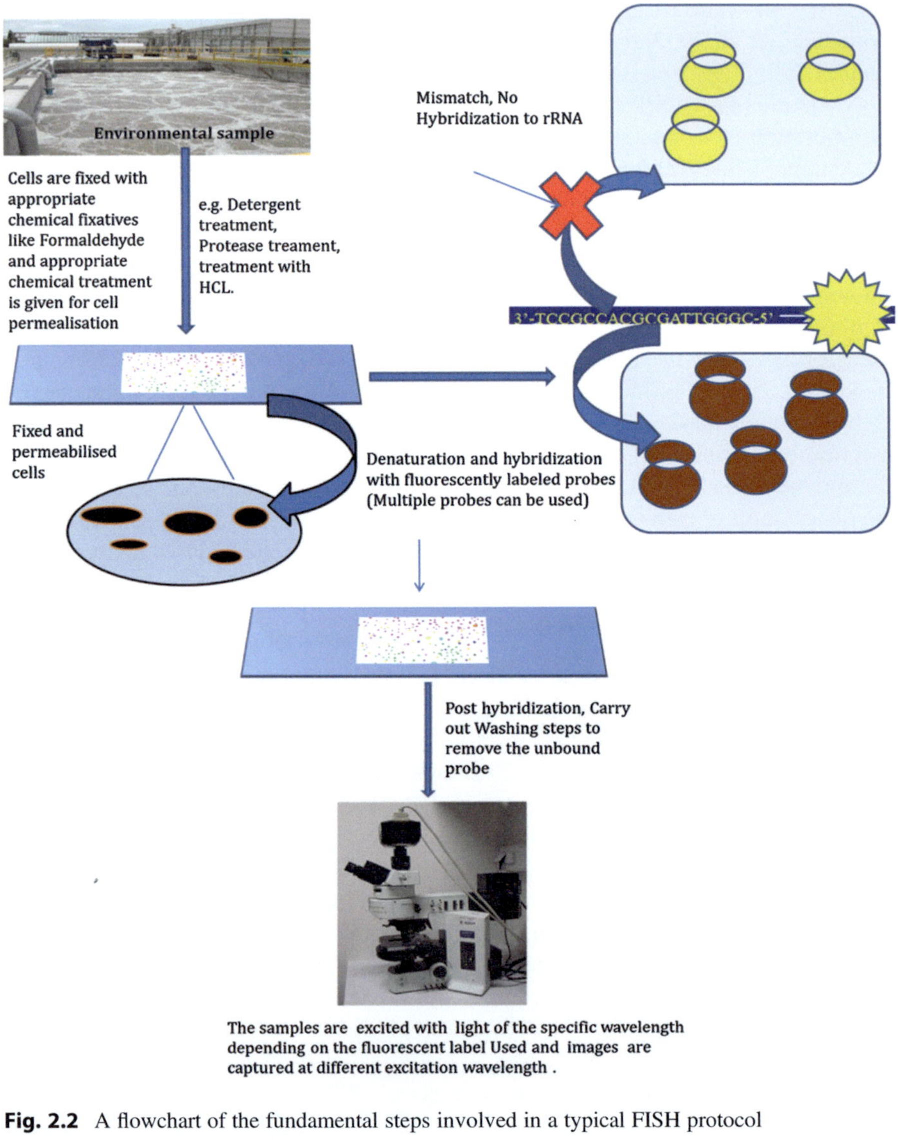

Fig. 2.2 A flowchart of the fundamental steps involved in a typical FISH protocol

samples. For instance, Zarda et al. (1997) used Cy3-labeled rRNA-targeted oligonucleotide probe "EUB338" to study the community structure of pristine forest soil of "Hau" (an aquic eutrochrept bulk soil) and revealed predominance of microbial members of α- and δ-subdivision of *Proteobacteria* and the *Planctomycetes* in targeted soil samples.

2.4.3 Microautoradiography

Microautoradiography involves the incubation of the environmental sample with a radiotracer for labeling of the microbial cell using radioisotopic compounds like soft beta emitters (^{3}H, ^{14}C, ^{33}P) or strong beta emitter like ^{32}P (Rogers et al. 2007). After incubation, the samples are fixed in paraformaldehyde or ethanol and washed to remove surplus radiotracer. The radiolabeled substrate is subsequently taken up by individual cells and releases beta decay particles that are used to develop the pattern using a radiation-sensitive emulsion silver halide emulsion. Excited silver ions will precipitate as metallic silver and will appear black grains after the development of the film. These beta decay particles reduce the silver ions in silver halide crystals to silver atoms generating silver grain clumps adjacent to radioactive cells. These silver grain clumps, so developed, can be easily seen clearly using transmission light microscopy/bright-field microscopy/phase-contrast microscopy/LSM (laser scanning microscopy). Microautoradiography is often combined with fluorescence in situ hybridization (MAR-FISH) which uses oligonucleotide probes for identification of the microorganisms in order to link the key metabolic features to the identity of the microorganism (Ouverney and Fuhrman 1999). The key determinant of a successful microautoradiography experiment is the exposure times which may range from a few hours to days and weeks depending on many factors which include the decay rate of the radioisotope used, substrate concentration labeled, isotope uptake rate, etc. Further, the cross-feeding of the labeled tracer may sometimes give erroneous interpretation. For instance, anticipated incorporation of labels specifically targeted for glucose-consuming bacteria, in certain conditions, can be incorporated and can cross-feed to other nontarget bacterial groups who cross-feed on labeled products produced by certain glucose-consuming bacteria and may give false-positive MAR signal. The microautoradiography, either alone or in conjunction with FISH, has been used to decipher the many aspects of soil microbial community structure and functions (Varró et al. 1986; Rogers et al. 2007; O'Donnell et al. 2007; Karbin et al. 2015).

2.4.4 DNA Microarrays

Deoxyribonucleic acid microarrays (also called DNA chips) were originally devised for the studies of differential gene expression in health-related issues, but their applications have also been extrapolated for the environmental studies like differential gene expression in response to environmental pollutants (Letowski et al. 2003). The forms of nucleic acid microarray chip, so-called GeoChips and PhyloChips, have been used to probe the microbial communities in various environmental samples (Asuming-Brempong 2012). In a DNA microarray experiment, expression of thousands of genes can be compared in two contrasting situations at the same time on the same chip. The basic principle of DNA microarrays involves hybridizations of two complementary, single-stranded regions of two DNA molecules, i.e., target and probe DNA molecules (usually either one is labeled with fluorescent dyes like Cy3 and Cy5), retrieved from two contrasting conditions. The hybridization events can be detected, owing to the labeling of the bound complementary target, by high-resolution scanning or imaging. A DNA microarray experiment is typically like the tradition nitrocellulose membrane hybridization experiment, but the probe and target relation in DNA microarray are reversed. In tradition hybridization experiments, the free probe (which is known) is usually labeled and the fixed target DNA (unknown) is not, while in a DNA microarray experiment, fixed DNA target (known sequences) is unlabeled and target DNA (unknown) is the labeled one. In microarray experiments, the testing substance is attached on the chip; hence some researcher prefers to call it a probe (by analogy to conventional hybridization experiments), as free DNA that we are querying is labeled (Zhou and Thompson 2002). In a typically DNA microarray experiment, the various experimental parameters like optimal probe concentration, its length, type of probe to use (i.e., oligonucleotide or amplicon), detection, and specificity limits to be expected have to be optimized, and further the interpretation and normalization of data to be reported are deciding factors and have to be considered while executing the experiment. The sensitivity of microarrays is always a critical factor. The oligonucleotide microarray displays higher sensitivity as reported by Small and co-workers, in their study on unpurified soil extracts. They were able to detect the *G. chapellei* SSU rRNA gene using ~0.5 μg of total RNA extracted from soils (Small et al. 2001).

2.4.5 Isotope Arrays

The isotope array is a very important technique in microbial ecology which is ideally suited to screen the microbial populations consuming particular substrates, even if no previous knowledge is available about such microbes. The technique basically involves incubating an environmental sample with a labeled substrate like ^{14}C-labeled substrate, followed by the extraction of RNA from the samples, which is then labeled with a fluorophore and used to hybridize with an oligonucleotide array that targets 16S rRNA gene of the bacteria of interest (Adamczyk et al. 2003). Post-hybridization, the array is scanned for fluorescence and incorporation of the radioactive isotope in order to determine which community members have incorporated the ^{14}C isotope into their RNA. The same probe which gives positive fluorescence signal and displays the incorporation of radiotracer can be retested by applying them with FISH-MAR with the same environmental samples (Hesselsoe et al. 2009). In their pioneer study based on combination of microarrays with the uptake of radioactive substrates, Adamczyk et al. (2003) were able to establish $^{14}CO_2$ fixation activities of the ammonia-oxidizing bacterial population within a complex activated sludge community, with identification and function of microorganisms in activated sludge. This method first time showed great isotope arrays as a technique that could simultaneously detect the composition and activity of specific populations of bacteria within wastewater treatment communities.

2.4.6 Quantitative PCR

Quantitative PCR, also called real-time polymerase chain reaction, is essentially based on analysis of specific DNA from an environmental sample which provides the estimate of number of genomes of a particular microbe per unit volume of the sample (Bustin et al. 2005; Smith and Osborn 2009). It also allows the quantification of the number of target genes in a community sample. Unlike conventional PCR, in the qPCR, a specific targeted DNA sequence is amplified and quantified simultaneously in real time, with progress of amplification reaction. The value so obtained corresponds to the number of genome in water sample, but does not give directly the number of cells. Several bacteria contain more than one copy of marker gene as rRNA genes and hence complicate the analysis. qPCR uses either intercalating fluorescent dyes such as SYBR Green or fluorescent probes (TaqMan probes, molecular beacons, scorpion probes, etc.) in order to measure the accumulation of PCR amplicons in real time as the amplification progresses. Several primers for the amplification of 16S rRNA, 5.8S rRNA, ITS gene, and functional genes (amoA, pmoA, norS, and dsrA) specific to ammonia oxidizers, methane oxidizers, and sulfate have been designed and employed in PCR-based quantification of soil bacterial and fungal microbial communities (Fierer et al. 2005; Foti et al. 2007). In a qPCR-based study, Kolb et al. (2003) reported the abundance of total methanotrophic population and specific groups of methanotrophs in a flooded rice field soil by qPCR assay of the pmoA genes.

2.4.7 Microbial Lipid Analysis

Phospholipid fatty acid (PLFA) analysis has been successfully used as a culture-independent method of assessing the structure of microbial communities in different environments (White et al. 2003; Keinänen et al. 2004; Goupil et al. 2015; Yao et al. 2016). The membranes of microorganisms have phospholipids which contain fatty acids; these phospholipid fatty acids are potentially useful signature molecules and are used to obtain microbial community fingerprints. Further, phospholipids are known to be present exclusively in cell membranes which are rapidly degraded following cell death and hence serve as important indicators of active microbial biomass. PLFA (phospholipid fatty acids) can be easily extracted from microbial cells in soil, and their fatty acid methyl esters (FAMEs) are accepted taxonomic discriminators for species identification which serve the basis for elucidating presence and abundance of a particular microbial group in community PLFA profiling. PLFA profile of a particular soil type can be used to link it with soil associated with particular cropping practices (Zelles et al. 1995) and can also be used to track pollution (Frostegard et al. 1993) and changes in soil quality (Reichardt et al. 1997; Bossio et al. 1998; Petersen et al. 1998). Bossio et al. (1998) studied two contrasted soil regimes, i.e., organically managed soils and soils receiving synthetic fertilizers and pesticides, and observed significantly different PLFA profiles in two regimes with greater diversity and abundance of aerobic bacteria, cyanobacteria, and methane-oxidizing bacteria in organically managed soil.

2.4.8 Cutting-Edge High-Throughput Sequencing for Community Structure Analysis

The high-throughput sequencing is definitely cutting the edge for community structure which is allowing us to get deeper insights into microbial community structure and their function in maintaining the major biogeochemical cycles of different ecosystems including soil, water, deep biosphere, etc. The sequencing technology has come a long way since the days of electrophoresis in the 1970s. With the advent of Maxam-Gilbert method also known as chemical sequencing method and Sanger chain termination method in 1977, scientists gained the ability to sequence DNA in a reliable manner. This method prevailed from the 1980s until the mid-2000. Because of its comparative ease and reliability, it is the method used in first-generation technology.

While these "first-generation" sequencing platforms are considered high throughput for their time, the short read massively parallel sequencing technique is a different approach that revolutionized sequence capabilities and launched "next generation" into genomic science. NGS is a term used to describe a number of various modern sequencing methods like Roche/454 GS, Solexa/Illumina, SOLiD, and Ion Torrent. It is because of NGS that researchers can now analyze thousands to tens of samples in a single year. NGS applies to genome sequencing/resequencing, transcriptomics (RNA-sequencing), ChIP sequencing, and epigenome characterization. The first NGS technology to be commercialized in 2005 is the pyrosequencing technique of 454 Life Sciences (now Roche) (Margulies et al. 2005). After that, Solexa/Illumina (2007), Sequencing by Oligonucleotide Ligation and Detection (SOLiD) by Applied Biosystems (2007), and Ion Torrent Personal Genome Machine (PGM) (2010) are commercialized (Bentley 2006; Metzker 2010). Besides, Qiagen-intelligent bio-systems sequencing, polony sequencing, and a single-molecule detection system (Helicos BioSciences) were also developed. Here, we discuss the five platforms, i.e., 454, Illumina, SOLiD, Ion Torrent, and PacBio, which are routinely used for high-throughput sequencing over the past decade.

2.4.8.1 Roche/454 GS

The method was developed by 454 Life Sciences (2004) and then overtaken by Roche Diagnostics. This sequencing method is based on the "sequencing by synthesis" principle. Here a single-stranded DNA is replicated to dsDNA by a polymerase enzyme. The enzyme sequentially added bases at the end of the DNA fragment. This process takes place inside the sequencing machine which contains many picoliter-volume wells. Each well contained single bead and sequencing enzymes. The sequencing process begins with the fragmentation of dsDNA into smaller fragments of DNA around 400–600 base pairs with the help of some restriction enzymes (Metzker 2010). Adapters (short sequence of DNA) are attached to the DNA fragments, and tiny resin beads are added to the mixture. The adaptor sequences complementary bind with template DNA which helps DNA fragments to bind directly to the beads. The DNA fragments are polymerized numerous times by polymerase chain reaction on each bead. Beads without sequence are filtered to remove, and the remaining DNA-containing beads are placed into wells on a sequencing plate for sequencing. Nucleotides are added to the wells in turns of one type of base at a time, i.e., A's followed by C's, G's, and T's. After single base incorporation into the nascent DNA, the chemical signals, i.e., light generated by luciferase enzyme, are converted into light that is recorded by CCD camera. The intensity of light varies proportionally with the consecutive number of nucleotides being analyzed (Mardis 2008). To decide the sequence of the original piece of DNA, this pattern of light intensity is plotted on the graph.

2.4.8.2 Solexa/Illumina

This technology is worked based on sequencing-by-synthesis method using reversible dye termination nucleotides. Along with DNA polymerase, all four fluorescent label nucleotides are added consecutively to the flow cell channels to sequence millions of clusters on the flow surface. The DNA is randomly fragmented (200–600 base pairs), and adapters are ligated to the end of the fragments. Unlabeled nucleotides and DNA polymerase are added to join the DNA strands which create "bridges" between dsDNA. Using heating, dsDNA is denatured into single-stranded DNA. The denaturation step leaves several millions of dense clusters of DNA that are produced in each flow channel. After that, the sequencing cycles are begun by adding primer, DNA polymerase, and four labeled reversible terminators (Mardis 2008). Using laser excitation, the emitted fluorescence from each cluster are captured and bases are identified. In Illumina sequencing, DNA sequence is analyzed base by base, making it a highly accurate method (Kozich et al. 2013).

2.4.8.3 Life Technologies SOLiD

Sequencing by Oligonucleotide Ligation and Detection (SOLiD) technology is developed by Life Technologies (2006). SOLiD is done by ligation and dual base encoding. The high accuracy of the SOLiD system allows analysis of samples across a wide range of applications. Two types of sample preparation method are present in SOLiD, i.e., fragment library (single DNA fragment) and mate-paired library (two DNA fragments). In both libraries, DNA is sheared into specific size and adapters are ligated to both the ends. There are millions of unique molecules in the library which represent the entire target sequence. In emulsion PCR, all the molecules associated with beads are clonally amplified. On the SOLiD system, the template-attached beads are combined with a universal sequencing primer, ligase, and a large pool of di-base probes. The di-base probes consist of a set of four fluorescently labeled nucleotides. The complementary probe hybridizes to the template and is ligated. After fluorescence is measured, the dye is cleaved from the 5′-phosphate group, and 5′ end is available for further reactions. This process is repeated several times (approximately seven cycles) to yield a 35 bp long read. The previously synthesized strand is removed. A new primer is hybridized, and base addition and the ligation cycles are repeated.

2.4.8.4 Benchtop Sequencers

The next-generation sequencing (NGS) portfolio was expanded by Life Technologies; this system also uses the technology of sequencing by synthesis. But this technology differs from the previous one; instead of fluorescence it measures the H^+ ion release during base incorporation. Chemical signals are directly transferred into digital information in Ion Torrent PGM machine. The first step in Ion Torrent PGM workflow is library construction. A library of DNA fragments is generated that is flanked by Ion Torrent adapters. The DNA fragments generated during library preparation are amplified onto Ion Sphere particles (beads). Amplification is accomplished by emulsion PCR, and Ion Sphere particles coated with the template are deposited in the chip wells. The template-loaded chip is placed on the Ion Torrent PGM sequencer. The data in Ion Torrent PGM runs through signal processing and base calling algorithms associated with individual reads. Individual bases are introduced one at a time and are incorporated by DNA polymerase. For each base incorporation, a proton is released that results in pH change. Every microwell of the PGM contains approximately one million copies of DNA. The pH change in every individual well is detected by ion sensor, which transforms the chemical changes into digital information. The chip records two bases if the voltage is doubled by the detection of two identical nucleotides. The generated output files of Ion Torrent system can be viewed and downloaded in sff, fastq, or sam/bam data formats.

2.4.8.5 Single-Molecule Real-Time Sequencing/Pacific Biosciences

Owing to single-cell sequencing technology, now it is possible to sequence a single stretch of DNA molecule extracted from a single or unique cell. Till date genome sequencing has mainly become possible from a large number of template DNA extracted from a culture of homogeneous bacterial population rather than a single cell. The technology does not need any prior PCR amplification of DNA fragments. The single-molecule real-time (SMRT™) DNA sequencing technology was industrialized by Pacific Biosciences. This technology enables a new paradigm in the genomic analysis by delivering longer reads and built-in flexibility. SMRT sequencing is built upon two key innovations, i.e., zero-mode waveguides (ZMWs), where light is illuminated at the bottom part of the well in which a template and DNA polymerase, phospholipid nucleotides complex is immobilized. SMRT sequencing relies on the principle of the single-molecule real-time sequencing performed in SMRT cells having millions of ZMWs. A ZMW is a cylindrical hole, hundreds of nanometers in diameter, performing a thin metal supported by a transparent substrate. The ZMW provides the world's smallest light detection volume. The DNA template is extended by the DNA polymerase with the fluorescently labeled dNTPs, and fluorescent tag which previously incorporated nucleotide is then cleaved off. The CCD camera captures this signal in a real-time PCR. This process runs in simultaneously/parallel up to thousands of ZMWs, which make up

Table 2.2 Pros and cons of all the next-generation sequencing platform

Technologies	Pros	Cons
Roche/454 GS	Low error rate	Medium/high start-up costs
		Must run at a large scale
	Medium read length (~400–600bp to 1 Kb)	Relatively high costs per base
	Cheaper and faster	It has difficulty in distinguishing the number of bases in a run of identical bases (such as AAAA)
Solexa/ Illumina	Low error rate	Must run at very large scale
	Lowest cost per base	Short read length (50–150 bp)
	Tons of data	Runs take multiple days
		High start-up costs
		De novo assembly difficult
SOLiD	Relatively accurate because each base is interrogated twice	Potential for error propagation across reads due to two-base encoding and sequential ligation
		High instrument cost
	High throughput and low cost per base	Short reads
		Relatively long run time
	Independent lanes can be run on 5500XL	
Ion Torrent PGM	Low start-up costs	Read lengths only ~100–200 bp so far
	Scalable (10–1000 Mb of data per run)	
	Low error rate	
	Fast runs (<3 hrs)	
	Medium/low cost per base	
SMRT/ PacBio	Can use single molecule as template	Medium/high cost per base
		High start-up costs
	Potential for very long reads (several 10Kb+)	

the SMRT cell. All the advantages and disadvantages of these next-generation techniques are presented in Table 2.2.

2.5 Conclusion and Future Scope

The constant improvement of molecular techniques, sequencing technology, and bioinformatics revolutionizes the field of soil microbial ecology, i.e., the identification of total microbial community structure, understanding the link between diversity and community structure and function. Almost a decade of the study in metagenomic techniques showed its ability to identify novel and rare unculturable organisms and their function in maintaining the major biogeochemical cycle and soil quality. Due to the higher throughput of data produced by "omics" studies, these approaches are gaining momentum, and more and more reported scientific studies are centered on these high-throughput approaches. However, the conventional techniques of microbial community analysis still remain important in the view that many findings of the high-throughput studies need to be validated and substantiated using conventional techniques like qPCR, FISH, and autoradiography. However, the integrated view of community dynamics can be generated by system biology approach which requires integration of data from various conventional and "omics" approaches, i.e., metagenomics, metatranscriptomics, metaproteomics, and meta-metabolomics. Since all different omics approaches provide biological information available at different levels, hence all the approaches have their own advantages and limitations discussed in-depth in further chapters. Therefore, complementing all the "omics" approaches with on another provides a better insight into functional and the physiological state of the microbial communities of a system. Therefore, use of multi-omics approaches for answering microbial ecology-related question and improvement of bioinformatics pipeline for the large data analysis and interpretation will be highly desirable in the future.

References

Adamczyk J, Hesselsoe M, Iversen N, Horn M, Lehner A, Nielsen PH et al (2003) The isotope array, a new tool that employs substrate-mediated labeling of rRNA for determination of microbial community structure and function. Appl Environ Microbiol 69(11):6875–6887

Adzitey F, Rusul G, Huda N, Cogan T, Corry J (2012) Prevalence, antibiotic resistance and RAPD typing of Campylobacter species isolated from ducks, their rearing and processing environments in Penang, Malaysia. Int J Food Microbiol 154(3):197–205

Antony CP, Kumaresan D, Ferrando L, Boden R, Moussard H, Scavino AF, Shouche YS, Murrell JC (2010) Active methylotrophs in the sediments of Lonar Lake, a saline and alkaline ecosystem formed by meteor impact. ISME J 4(11):1470–1480

Asuming-Brempong S (2012) Microarray technology and its applicability in soil science–a short review. Open J Soil Sci 2(3):333

Barakat MA (2011) New trends in removing heavy metals from industrial wastewater. Arab J Chem 4:361–377

Beaver CL, Williams AE, Atekwana EA, Mewafy FM, Abdel Aal G, Slater LD, Rossbach S (2016) Microbial communities associated with zones of elevated magnetic susceptibility in hydrocarbon-contaminated sediments. Geomicrobiol J 33(5):441–452

Bentley DR (2006) Whole-genome re-sequencing. Curr Opin Genet Dev 16(6):545–552

Błaszczyk D, Bednarek I, Machnik G, Sypniewski D, Sołtysik D, Loch T, Gałka S (2011) Amplified ribosomal dna restriction analysis (ARDRA) as a screening method for normal and bulking activated sludge sample differentiation. Pol J Environ Stud 20:29–36

Bossio DA, Scow KM, Gunapala N, Graham KJ (1998) Determinants of soil microbial communities: effects of agricultural management, season, and soil type on phospholipid fatty acid profiles. Microb Ecol 36:1–12

Bustin SA, Benes V, Nolan T, Pfaffl MW (2005) Quantitative real-time RT-PCR – a perspective. J Mol Endocrinol 34:597–601

Ciesielski S, Bułkowska K, Dabrowska D, Kaczmarczyk D, Kowal P, Możejko J (2013) Ribosomal intergenic spacer analysis as a tool for monitoring methanogenic archaea changes in an anaerobic digester. Curr Microbiol 67(2):240–248

Cleary DF, Smalla K, Mendonça-Hagler LC, Gomes NC (2012) Assessment of variation in bacterial composition among microhabitats in a mangrove environment using DGGE fingerprints and barcoded pyrosequencing. PLoS One 7(1):e29380

Cummings DE, Snoeyenbos-West OL, Newby DT, Niggemyer AM, Lovley DR, Achenbach LA, Rosenzweig RF (2003) Diversity of Geobacteraceae species inhabiting metal-polluted freshwater lake sediments ascertained by 16S rDNA analyses. Microb Ecol 46(2):257–269

DeRito CM, Pumphrey GM, Madsen EL (2005) Use of field-based stable isotope probing to identify adapted populations and track carbon flow through a phenol-degrading soil microbial community. Appl Environ Microbiol 71(12):7858–7865

Dodd JC, Boddington CL, Rodriguez A, Gonzalez-Chavez C, Mansur I (2000) Mycelium of arbuscular mycorrhizal fungi (AMF) from different genera: form, function and detection. Plant Soil 226(2):131–151

Dorn-In S, Hölzel CS, Janke T, Schwaiger K, Balsliemke J, Bauer J (2013) PCR-SSCP-based reconstruction of the original fungal flora of heat-processed meat products. Int J Food Microbiol 162(1):71–81

Dunbar J, Barns SM, Ticknor LO, Kuske CR (2002) Empirical and theoretical bacterial diversity in four Arizona soils. Appl Environ Microbiol 68(6):3035–3045

Dunford EA, Neufeld JD (2010) DNA stable-isotope probing (DNA-SIP). JoVE (J Vis Exp) 42: e2027–e2027

Feligini M, Panelli S, Sacchi R, Ghitti M, Capelli E (2015) Tracing the origin of raw milk from farm by using Automated Ribosomal Intergenic Spacer Analysis (ARISA) fingerprinting of microbiota. Food Control 50:51–56

Felske A, Akkermans AD, De Vos WM (1998) In situ detection of an uncultured predominant Bacillus in Dutch grassland soils. Appl Environ Microbiol 64(11):4588–4590

Fierer N, Jackson JA, Vilgalys R, Jackson RB (2005) Assessment of soil microbial community structure by use of taxon-specific quantitative PCR assays. Appl Environ Microbiol 71:4117–4120

Filion M, St-Arnaud M, Fortin JA (1999) Direct interaction between the arbuscular mycorrhizal fungus Glomus intraradices and different rhizosphere microorganisms. New Phytol 141 (3):525–533

Fischer SG, Lerman LS (1980) Separation of random fragments of DNA according to properties of their sequences. Proc Natl Acad Sci 77(8):4420–4424

Fisher MM, Triplett EW (1999) Automated approach for ribosomal intergenic spacer analysis of microbial diversity and its application to freshwater bacterial communities. Appl Environ Microbiol 65(10):4630–4636

Foti M, Sorokin DY, Lomans B, Mussman M, Zacharova EE, Pimenov NV, Kuenen JG, Muyzer G (2007) Diversity, activity, and abundance of sulfate-reducing bacteria in saline and hypersaline soda lakes. Appl Environ Microbiol 73:2093–3000

Franklin RB, Taylor DR, Mills AL (1999) Characterization of microbial communities using randomly amplified polymorphic DNA (RAPD). J Microbiol Methods 35(3):225–235

Frostegard A, Tunlid A, Baath E (1993) Phospholipid fatty acid composition, biomass and activity of microbial communities from two soil types experimentally exposed to different heavy metals. Appl Environ Microbiol 59:3605–3617

Garbeva P, Van Veen JA, Van Elsas JD (2004) Microbial diversity in soil: selection of microbial populations by plant and soil type and implications for disease suppressiveness. Annu Rev Phytopathol 42:243–270

George E, Marschner H, Jakobsen I (1995) Role of arbuscular mycorrhizal fungi in uptake of phosphorus and nitrogen from soil. Crit Rev Biotechnol 15(3–4):257–270

Goszczynski DE (2007) Single-strand conformation polymorphism (SSCP), cloning and sequencing reveal a close association between related molecular variants of Grapevine virus A (GVA) and Shiraz disease in South Africa. Plant Pathol 56(5):755–762

Goszczynski DE, Jooste AEC (2015) The application of single-strand conformation polymorphism (SSCP) technique for the analysis of molecular heterogeneity of grapevine virus A. VITIS-J Grapevine Res 41(2):77

Goupil K, Nkongolo KK, Nasserulla S (2015) Characterization of fungal communities in limed and unlimed lands contaminated with metals: phospholipid fatty acid (PLFA) analysis and soil respiration. Am J Biochem Biotechnol 11(2):45

Gulitz A, Stadie J, Ehrmann MA, Ludwig W, Vogel RF (2013) Comparative phylobiomic analysis of the bacterial community of water kefir by 16S rRNA gene amplicon sequencing and ARDRA analysis. J Appl Microbiol 114(4):1082–1091

Hesselsoe M, Füreder S, Schloter M, Bodrossy L, Iversen N, Roslev P et al (2009) Isotope array analysis of Rhodocyclales uncovers functional redundancy and versatility in an activated sludge. ISME J 3(12):1349–1364

Heyndrickx M, Vauterin L, Vandamme P, Kersters K, De Vos P (1996) Applicability of combined amplified ribosomal DNA restriction analysis (ARDRA) patterns in bacterial phylogeny and taxonomy. J Microbiol Methods 26(3):247–259

Hill GT, Mitkowski NA, Aldrich-Wolfe L, Emele LR, Jurkonie DD, Ficke A et al (2000) Methods for assessing the composition and diversity of soil microbial communities. Appl Soil Ecol 15 (1):25–36

Hugenholtz P (2002) Exploring prokaryotic diversity in the genomic era. Genome Biol 3:1. reviews0003-1

Ivone VM, Conceição E (2013) Bacterial diversity from the source to the tap: a comparative study based on 16S rRNA gene-DGGE and culture-dependent methods. FEMS Microbiol Ecol 83 (2):361–374

Jehmlich N, Schmidt F, Hartwich M, von Bergen M, Richnow HH, Vogt C (2008) Incorporation of carbon and nitrogen atoms into proteins measured by protein-based stable isotope probing (Protein-SIP). Rapid Commun Mass Spectrom 22:2889–2897

Jehmlich N, Kopinke FD, Lenhard S, Herbst FA, Seifert J, Lissner U, Vo¨lker U, Schmidt F, von Bergen M (2012) Sulphur-36S stable isotope labeling of amino acids for quantification (SULAQ). Proteomics 12:37–42

Karbin S, Guillet C, Kammann CI, Niklaus PA (2015) Effects of long-term CO_2 enrichment on soil-atmosphere CH 4 fluxes and the spatial micro-distribution of methanotrophic bacteria. PLoS One 10(7):e0131665

Keinänen MM, Martikainen PJ, Kontro MH (2004) Microbial community structure and biomass in developing drinking water biofilms. Can J Microbiol 50(3):183–191

Key KC, Sublette KL, Duncan K, Mackay DM, Scow KM, Ogles D (2013) Using DNA-Stable isotope probing to identify MTBE- and TBA-degrading microorganisms in contaminated groundwater. Ground Water Monit Remediat 33:57–68

Kindaichi T, Awata T, Mugimoto Y, Rathnayake RM, Kasahara S, Satoh H (2016) Effects of organic matter in livestock manure digester liquid on microbial community structure and in situ activity of anammox granules. Chemosphere 159:300–307

Kolb S, Knief C, Stubner S, Conrad R (2003) Quantitative detection of methanotrophs in soil by novel pmoA-targeted real-time PCR assays. Appl Environ Microbiol 69:2423–2429

Kozich JJ, Westcott SL, Baxter NT, Highlander SK, Schloss PD (2013) Development of a dual-index sequencing strategy and curation pipeline for analyzing amplicon sequence data on the MiSeq Illumina sequencing platform. Appl Environ Microbiol 79(17):5112–5120

Letowski J, Brousseau R, Masson L (2003) DNA microarray applications in environmental microbiology. Anal Lett 36(15):3165–3184

Liou JS-C, DeRito CM, Madsen EL (2008) Field-based and laboratory stable isotope probing surveys of the identities of both aerobic and anaerobic benzene-metabolizing microorganisms in freshwater sediment. Environ Microbiol 10:1964–1977

Liu WT, Marsh TL, Cheng H, Forney LJ (1997) Characterization of microbial diversity by determining terminal restriction fragment length polymorphisms of genes encoding 16S rRNA. Appl Environ Microbiol 63(11):4516–4522

Lu Y, Conrad R (2005) In situ stable isotope probing of methanogenic archaea in the rice rhizosphere. Science 309(5737):1088–1090

Lueders T, Wagner B, Claus P, Friedrich MW (2004) Stable isotope probing of rRNA and DNA reveals a dynamic methylotroph community and trophic interactions with fungi and protozoa in oxic rice field soil. Environ Microbiol 6:60–72

MacNaughton SJ, Stephen JR, Venosa AD, Davis GA, Chang YJ, White DC (1999) Microbial population changes during bioremediation of an experimental oil spill. Appl Environ Microbiol 65(8):3566–3574

Malik S, Beer M, Megharaj M, Naidu R (2008) The use of molecular techniques to characterize the microbial communities in contaminated soil and water. Environ Int 34(2):265–276

Manefield M, Whiteley AS, Griffiths RI, Bailey MJ (2002) RNA stable isotope probing, a novel means of linking microbial community function to phylogeny. Appl Environ Microbiol 68 (11):5367–e5373

Mardis ER (2008) Next-generation DNA sequencing methods. Annu Rev Genomics Hum Genet 9:387–402

Margulies M, Egholm M, Altman WE, Attiya S, Bader JS, Bemben LA et al (2005) Genome sequencing in microfabricated high-density picolitre reactors. Nature 437(7057):376–380

McGarry JW, Ortiz PL, Hodgkinson JE, Goreish I, Williams DJ (2013). PCR-based differentiation of Fasciola species (Trematoda: Fasciolidae), using primers based on RAPD-derived sequences. Ann Trop Med Parasitol

Metzker ML (2010) Sequencing technologies-the next generation. Nat Rev Genet 11(1):31–46

Molin J, Molin S (1997) CASE: complex adaptive systems ecology. In: Advances in microbial ecology. Springer US, Boston, pp 27–79

Müller H, Bosch J, Griebler C, Damgaard LR, Nielsen LP, Lueders T, Meckenstock RU (2016) Long-distance electron transfer by cable bacteria in aquifer sediments. ISME J 10:2010

Muyzer G, De Waal EC, Uitterlinden AG (1993) Profiling of complex microbial population by DGGE analysis of polymerase chain reaction amplified genes encoding for 16S rRNA. Appl Environ Microbiol 62:2676–2680

Neufeld JD, Vohra J, Dumont MG, Lueders T, Manefield M, Friedrich MW, Murrell JC (2007) DNA stable-isotope probing. Nat Protoc 2(4):860–866

Neufeld JD, Chen Y, Dumont MG, Murrell JC (2008) Marine methylotrophs revealed by stable-isotope probing, multiple displacement amplification and metagenomics. Environ Microbiol 10:1526–1535

O'Donnell GA et al (2007) Visualization, modelling and prediction in soil microbiology. Nat Rev Microbiol 5:9689–9699

Orita M, Suzuki Y, Sekiya T, Hayashi K (1989) Rapid and sensitive detection of point mutations and DNA polymorphisms using the polymerase chain reaction. Genomics 5(4):874–879

Osborn AM, Moore ER, Timmis KN (2000) An evaluation of terminal-restriction fragment length polymorphism (T-RFLP) analysis for the study of microbial community structure and dynamics. Environ Microbiol 2(1):39–50

Ouverney CC, Fuhrman JA (1999) Combined microautoradiography-16S rRNA probe technique for determination of radioisotope uptake by specific microbial cell types in situ. Appl Environ Microbiol 65(4):1746–1752

Paes F, Liu X, Mattes TE, Cupples AM (2015) Elucidating carbon uptake from vinyl chloride using stable isotope probing and Illumina sequencing. Appl Microbiol Biotechnol 99:7735–7743

Pascual J, Blanco S, García-López M, García-Salamanca A, Bursakov SA, Genilloud O et al (2016) Assessing bacterial diversity in the rhizosphere of thymus zygis growing in the Sierra Nevada National Park (Spain) through culture-dependent and independent approaches. PLoS One 1(1): e0146558

Petersen SO, Debosz K, Schjonning P, Christensen BT, Elmholt S (1998) Phospholipid fatty acid profiles and C availability in wet-stable macro-aggregates from conventionally and organically farmed soils. Geoderma 78:181–196

Radajewski S, Ineson P, Parekh NR, Murrell JC (2000) Stable-isotope probing as a tool in microbial ecology. Nature 403:646–649

Radajewski S, Webster G, Reay DS, Morris SA, Ineson P, Nedwell DB, Prosser JI, Murrell JC (2002) Identification of active methylotroph populations in an acidic forest soil by stable-isotope probing. Microbiology 148:2331–2342
Ralebitso TK, Röling WF, Braster M, Senior E, van Verseveld HW (2000) 16S rDNA-based characterization of BTX-catabolizing microbial associations isolated from a South African sandy soil. Biodegradation 11(6):351–357
Rangel-Castro JI, Killham K, Ostle N, Nicol GW, Anderson IC, Scrimgeour CM, Ineson P, Meharg AA, Prosser JI (2005) Stable isotope probing analysis of the influence of liming on root exudate utilization by soil microorganisms. Environ Microbiol 7:828–838
Reichardt W, Mascarina G, Padre B, Doll J (1997) Microbial communities of continuously cropped, irrigated rice fields. Appl Environ Microbiol 63:233–238
Rickwood D (1992) Centrifugal methods for characterizing macromolecules and their interactions. In: Rickwood D (ed) Preparative centrifugation: a practical approach. Oxford University Press, Oxford, pp 143–186
Rogers SW, Thomas BM, Say KO (2007) Fluorescent in situ hybridization and micro-autoradiography applied to ecophysiology in soil. Soil SciSoc Am J 71(2):620–631
Schloss PD, Handelsman J (2004) Status of the microbial census. Microbiol Mol Biol Rev 68:686–691
Schwieger F, Tebbe CC (1998) A new approach to utilize PCR–single-strand-conformation polymorphism for 16S rRNA gene-based microbial community analysis. Appl Environ Microbiol 64 (12):4870–4876
Seifert J, Taubert M, Jehmlich N, Schmidt F, Völker U, Vogt C et al (2012) Protein-based stable isotope probing (protein-SIP) in functional metaproteomics. Mass Spectrom Rev 31(6):683–697
Sheffield VC, Beck JS, Kwitek AE, Sandstrom DW, Stone EM (1993) The sensitivity of single-strand conformation polymorphism analysis for the detection of single base substitutions. Genomics 16(2):325–332
Shehata AI (2012) Molecular identification of probiotics Lactobacillus strain isolates by: amplified ribosomal DNA restriction analysis (ARDRA). Sci J Microbiol 6:3034
Singh BK, Campbell CD, Sorenson SJ, Zhou J (2009) Soil genomics. Nat Rev Microbiol 7 (10):756–756
Small J, Call DR, Brockman FJ, Straub TM, Chandler DP (2001) Direct detection of 16S rRNA in soil extracts by using oligonucleotide microarrays. Appl Environ Microbiol 67:4708–4716
Smith KP, Goodman RM (1999) Host variation for interactions with beneficial plant-associated microbes. Annu Rev Phytopathol 37(1):473–491
Smith CJ, Osborn AM (2009) Advantages and limitations of quantitative PCR (Q-PCR)-based approaches in microbial ecology. FEMS Microbiol Ecol 67:6–20
Srivastava J, Lambert J, Vietmeyer N (1996) Medicinal plants: An expanding role in development, vol Vol. 320. World Bank Publications, Washington, DC
Taubert M, Baumann S, von Bergen M, Seifert J (2011) Exploring the limits of robust detection of incorporation of ^{13}C by mass spectrometry in protein-based stable isotope probing (protein-SIP). Anal Bioanal Chem 401:1975–1982
Tiedje JM, Asuming-Brempong S, Nüsslein K, Marsh TL, Flynn SJ (1999) Opening the black box of soil microbial diversity. Appl Soil Ecol 13(2):109–122
Timonen S, Finlay RD, Olsson S, Söderström B (1996) Dynamics of phosphorus translocation in intact ectomycorrhizal systems: non-destructive monitoring using a β-scanner. FEMS Microbiol Ecol 19(3):171–180
Uhlik O, Leewis MC, Strejcek M, Musilova L, Mackova M, Leigh MB, Macek T (2013) Stable isotope probing in the metagenomics era: a bridge towards improved bioremediation. Biotechnol Adv 31(2):154–165
Vaneechoutte M, De Beenhouwer H, Claeys G, Verschraegen GERDA, De Rouck A, Paepe N et al (1993) Identification of Mycobacterium species by using amplified ribosomal DNA restriction analysis. J Clin Microbiol 31(8):2061–2065

Vaneechoutte M, Riegel P, De Briel D, Monteil H, Verschraegen GERDA, De Rouck A, Claeys G (1995) Evaluation of the applicability of amplified rDNA-restriction analysis (ARDRA) to identification of species of the genus *Corynebacterium*. Res Microbiol 146(8):633–641

Varró T, Gelencser J, Somogyi G (1986) Study of transport processes in soils and plants by microautoradiographic and radioabsorption methods. Acta Biochim Biophys Hung 22(1):31–43

Vogt C, Lueders T, Richnow HH, Krüger M, von Bergen M, Seifert J (2016) Stable isotope probing approaches to study anaerobic hydrocarbon degradation and degraders. J Mol Microbiol Biotechnol 26(1–3):195–210

Wald J, Hroudova M, Jansa J, Vrchotova B, Macek T, Uhlik O (2015) Pseudomonads rule degradation of polyaromatic hydrocarbons in aerated sediment. Front Microbiol 6:1268

Wall DH, Virginia RA (1999) Controls on soil biodiversity: insights from extreme environments. Appl Soil Ecol 13(2):137–150

Watanabe K, Kodama Y, Harayama S (2001) Design and evaluation of PCR primers to amplify bacterial 16S ribosomal DNA fragments used for community fingerprinting. J Microbiol Methods 44(3):253–262

White DC, Gouffon JS, Peacock AD, Geyer R, Biernacki A, Davis GA, Pryor M, Tabacco MB, Sublette KL (2003) Forensic analysis by comprehensive rapid detection of pathogens and contamination concentrated in biofilms in drinking water systems for water resource protection and management. Environ Forensic 4:63–74

Whiteley AS et al (2007) RNA stable-isotope probing. Nat Protoc 2(4):838–844

Whitman WB, Coleman DC, Wiebe WJ (1998) Prokaryotes: the unseen majority. Proc Natl Acad 95(12):6578–6583

Williams JG, Kubelik AR, Livak KJ, Rafalski JA, Tingey SV (1990) DNA polymorphisms amplified by arbitrary primers are useful as genetic markers. Nucleic Acids Res 18 (22):6531–6535

Wright SF, Upadhyaya A (1998) A survey of soils for aggregate stability and glomalin, a glycoprotein produced by hyphae of arbuscular mycorrhizal fungi. Plant Soil 198(1):97–107

Yao H, He ZL, Wilson M, Campbell CD (2000) Microbial biomass and community structure in a sequence of soils with increasing fertility and changing land use. Microb Ecol 40(3):223–237

Yao Z, Xing J, Gu H, Wang H, Wu J, Xu J, Brookes PC (2016) Development of microbial community structure in vegetable-growing soils from open-field to plastic-greenhouse cultivation based on the PLFA analysis. J Soils Sediments:16, 1–19

Zarda B, Hahn D, Chatzinotas A, Schonhuber W, Neef A, Amann RI, Zeyer J (1997) Analysis of bacterial community structure in bulk soil by in situ hybridization. Arch Microbiol 168:185–192

Zelles L, Rackwitz R, Bai QY, Beck T, Beese F (1995) Discrimination of microbial diversity by fatty acid profiles of phospholipids and lipopolysaccharides in differently cultivated soils. Plant Soil 170:115–122

Zhang Y, Deng W, Xie X, Jiao N (2016) Differential Incorporation of Carbon Substrates among Microbial Populations Identified by Field-Based, DNA Stable-Isotope Probing in South China Sea. PLoS One 11(6):e0157178

Zhou J, Thompson DK (2002) Challenges in applying microarrays to environmental studies. Curr Opin Biotechnol 13:204–207

'Omics' Tools in Soil Microbiology: The State of the Art

3

Rimi Biswas and Angana Sarkar

Abstract

The soil being the most heterogeneous substance hosts the dynamic environments for diverse microorganisms. Traditional techniques are limited to explore only few portion of massive unknown soil microbial world due to their well-known biasness in detecting microbial genetics and functional diversity. With this respect, omics targets the powerful genomics, metagenomics, transcriptomics, proteomics and metabolomic tools to explore the vast microbial community, new biomolecules and novel pathways. It helps to better understand the toxicity mechanisms, predicts the risks associated with environmental toxicity and aids in bioprospecting of value-added products. These new approaches will be useful to establish the linkage between structure and function of soil microbial community and help to get better insight of the ecological processes in the environment with special emphasis on plant-microbe ecosystems. The present chapter will give an overview of the application of the advanced molecular tools as well as their potentials and limitations in studying the soil microbial ecology.

Keywords

Omics · Soil · Bacteria · Microbiology · Ecology

R. Biswas · A. Sarkar (✉)
Department of Biotechnology and Medical Engineering, National Institute of Technology Rourkela, Rourkela, Odisha, India
e-mail: sarkara@nitrkl.ac.in

T. K. Adhya et al. (eds.), *Advances in Soil Microbiology: Recent Trends and Future Prospects*, Microorganisms for Sustainability 3,
https://doi.org/10.1007/978-981-10-6178-3_3

3.1 Introduction: Basic Concepts of 'Omics': Beyond Sequencing

3.1.1 Overview

Soil is the crucial part of our ecosystem having enormous potential of supporting microbial growth, nutrient enhancement and contaminant degradation if sustained naturally in its pristine condition. Soil microbial communities play a crucial role in plant-microbe interaction by cycling the essential nutrients through mineralization and by decomposition of organic matter. These communities influence the soil's biogeochemical cycle by altering the nutrient availability through solubilization, chelation and oxidation/reduction processes. The soil being the most heterogeneous has a dynamic environment for the growth of different organisms including prokaryotes as well as archaea, viruses, algae and fungi. Hence, it hosts a microbial population of immense diversity which leads to a high degree of competition, predation and parasitism among the microorganisms. On the other hand, the physical and chemical components of the natural habitat may sometimes play as the limiting factor as the soil particle size is generally inversely proportional to the number and diversity of prokaryotes in the soil (Sessitsch et al. 2001). The temperature, pH, organic matter and the moisture content of the soil also play an important role in shaping the microbial communities of each species (Hassink et al. 1993). Appropriate understanding of the soil ecosystem demands clarity of the three most fundamental questions: (i) who are they? (ii) what are they doing? and (iii) how are they interacting with the soil ecosystem? The genome-enabled modern molecular tools aim to answer these central questions with the help of its five fingers which include genomics, metagenomics, metaproteomics, metatranscriptomics and metabolomics. A combined effort of these tools is bringing in the new epoch of unrevealed microbial world. Rapid advancement in the field of 'Omics' technologies has led to an increased level of understanding the soil microbial community and its associated counterparts. Metagenomics has an unprecedented upper hand in comparison with the traditional culture techniques as it can detect a wider and a specific range of microorganisms including the unculturable ones. Therefore, molecular methods together with next-generation sequencing techniques can potentially provide a more reliable, effective and environmental friendly technology in detecting microbial cultures and their related activities in the soil. Different combinations of 'Omics' tools along with various bioinformatic approaches increase the study of integrated activity patterns and the growth of the soil microorganisms among themselves as well as the metabolism of plants associated with them (Mocalli and Benedetti 2010).

The term 'Omics' refers to the quantification or characterization of a set of molecules which are generally biological in nature to detect the function, structure, physiology and the molecular mechanisms of a set of organisms (Yun-Feng 2013). Metagenomics aims at the study of such organisms whose genetic materials can directly be assessed from the samples recovered from the environment, whereas proteomics and transcriptomics are the study of the structure, function and components of different proteins and RNA molecules. Nowadays most of the soils are contaminated due to the deposition of heavy metals in variable concentrations and indefinite compositions. The role of soil microbes and their mechanism of action to restore the contaminated sites has since been the foremost factor for the bioremediation of polluted soils. Microarray-based metagenomics and sequencing is the most saturated approach among all the metagenomic studies which provides a clear and transparent information regarding other 'Omics' technologies (Yun Feng 2013). Microbes serve as an important source of naturally occurring enzymes and secondary metabolites such as antibiotics. Therefore, to overcome the difficulty in cultivation for the majority of the microbes, intensive efforts are being made by increasing the time of incubation and properly simulating the naturally occurring environment along with the encapsulation of microbes which allows them to form physically separate groups and continue their molecular exchanges with the environment.

3.1.2 Application of 'Omics' in Soil System

Omics-based data comprehensively delivers a snapshot of the expressions of relevant genes and proteins as well as of the metabolite pattern which overall can provide a deeper insight into activities of the organisms with respect to its interaction with the soil microenvironment. Applications of Omics technologies in the soil system resulted in the emergence of a new arena. Omics technologies comprising of genomics, metagenomics, transcriptomics, proteomics and metabolomics provide ample in-depth information for a detailed understanding of the soil microbial community, their functional genetic regulatory factors, mode of toxicity and mechanisms of their interactions (Garbeva et al. 2004) (Fig. 3.1). Application of Omics is still in its early stage of development and hence needs some continuous improvement for better understanding of soil biota and their biogeochemical cycling (Fig. 3.1 and Table 3.1).

The diversity and activity of microorganisms in the Earth's biosphere is revealed in microbial ecology. In the last two decades, the application of advance genomic tools has revolutionized microbial ecological studies drastically and expanded our assessment on the previously unexplored microbial world. In the last few years, researchers have focused on the projects related to environmental effect monitoring using different culture-dependent and culture-independent molecular approaches (Bastida et al. 2009). Furthermore, metagenomics techniques have allowed rapid exploration of bacterial community structure and their functions in diverse soil ecosystem (Riesenfeld et al. 2004). The application of modern molecular tools not only refers to studies involving just single macromolecules, but actually involves the study of complete cellular pathways of entire organisms (Altschul et al. 1990). Fortunately, worldwide genome sequencing projects are producing incredible amount of sequencing data for analysing soil microbial community and their activities, but still the data generated is not sufficient enough to fill the knowledge gap for analysing enormous soil microbial diversity.

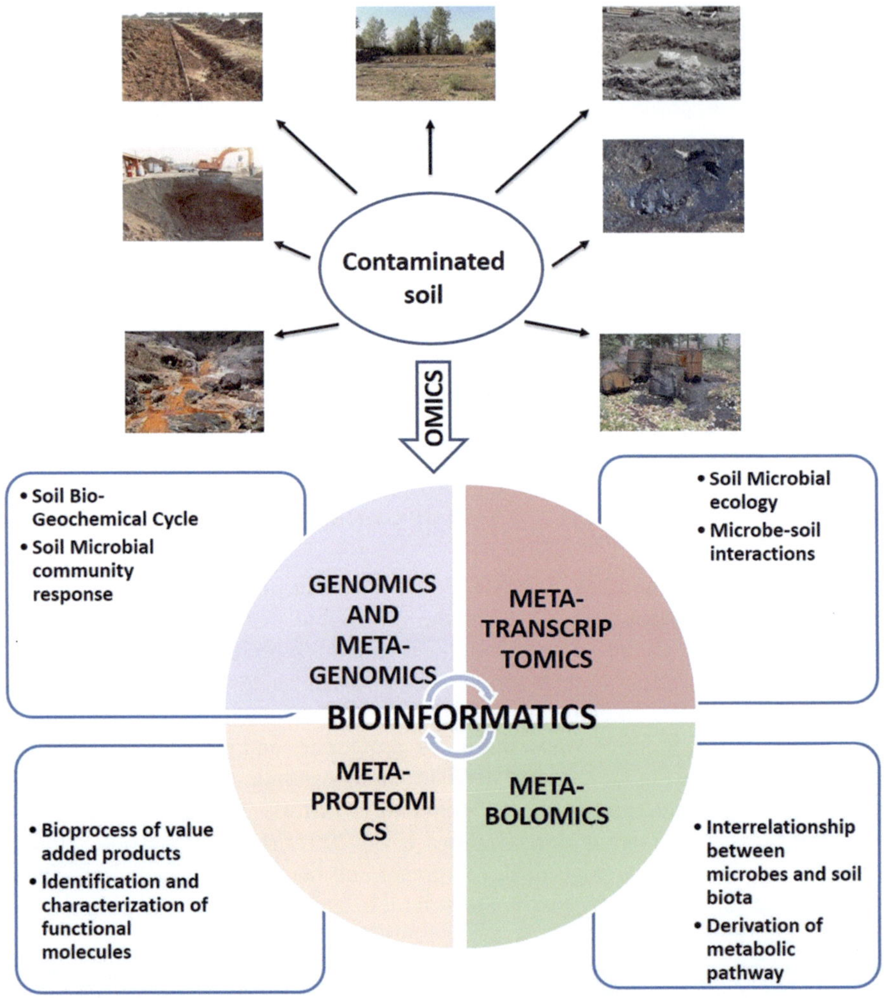

Fig. 3.1 Application of Omics tools in soil ecology: a conceptual framework

Nowadays bioremediation using microbes has become one of the most effective, cost-competitive and eco-friendly way of reducing contaminants, anthropogens and toxins from the polluted sites. A better assessment of the soil functionality along with its physiological characterization can be done using metagenomics, transcriptomics and proteomics in combination with various bioinformatic approaches. Rapid advances in the field of second-generation or next-generation sequencing have led to the sole dependency of metagenomics on these techniques. Based on pyrosequencing and sequencing by synthesis technology, a genome sequencer called Roche 454 was the first commercially available sequencer in this field. Presently junior system and FS FLX system are the versions available for

Table 3.1 Study of diverse soil organisms/microbial community using different advanced molecular tools

Tools	Analysis	Isolation source	References
Genomics	Colonization of wheat roots by a *Pantoea agglomerans* strain	Moroccan vertisol	Amellal et al. (1998)
	Soybean rhizobia: Genetic characterization	Alto Parana, Itapua, Paraguay	Chen et al. (2000)
	Genetic characterization of atrazine-degrading *Pseudaminobacter*	French and Canadian agricultural soils	Topp et al. (2000)
	Detection of archaea among other uncultivated lineages	Garden soil	Quaiser et al. (2002)
	Evidence of acidobacteria using environmental genomics	Calcareous grassland soil	Quaiser et al. (2003)
	Genomic studies of archaea using large insert DNA libraries from soil	Sandy and forest soil	Treusch et al. (2004)
	Study of uncultivated archaea using genomic studies	Porous soil	Schleper et al. (2005)
	Phylogenetic and biogeographic diversity of Cyanidiales	Yellowstone National Park, Japan and New Zealand	Toplin et al. (2008)
	Draft genome sequence of a bacteria tolerant to heavy metals: *Caulobacter*	Subsurface sediments from Oak Ridge, TN	Utturkar et al. (2013)
	Application of single cell genomic analysis for the enrichment of root endophytic bacteria from *Populus deltoides*	Fertile agricultural soil	Utturkar et al. (2016)
	Breeding to root system using genomic approach to enhance rice production	Enriched fertile soil	Uga (2017)
	Mining of lignocellulosic-degrading enzymes from semiarid soil	Semiarid soil	Junior et al. (2017)
Metagenomics	Study of soil microbial functions and diversity	Temperate forest soil	Torvisk and Ovreas (2002)
	Indigo and indirubin production in *Escherichia coli* by characterization of a clone of forest soil metagenome	Jindong Valley, Korea	Lim et al. (2005)
	Study of the genetic diversity of archaea, bacteria, viruses and fungi in soil	Individual soil sample	Fierer et al. (2007)
	Expression of antifungal activity in *Escherichia coli* using metagenome gene cluster in forest soil	Forest soil	Chung et al. (2008)

(continued)

Table 3.1 (continued)

Tools	Analysis	Isolation source	References
	Isolation of metagenomic library from Antarctic topsoil followed by	Antarctic topsoil	Cieśliński et al. (2009)
	Escherichia coli carrying soil metagenomic genes produces intermediates of porphyrin	Rice paddy soil	Kim et al. (2009)
	Identification of lipolytic gene families using metagenomic libraries from soil samples	Forest and grassland soil, Germany	Nacke et al. (2011)
	Construction of metabolic pathways associated with C, N and S cycling	Mangrove forests in Brazil	Andreote et al. (2012)
	Soil metagenome studies	Nunavut, Canada	Yergeau et al. (2012)
	Comparison of gene content by assembly of DNA sequences into larger fragments	Minnesota farm soil	Myrold et al. (2013)
	Identifying soil diversity using complex metagenomes	Farm soil	Howe et al. (2014)
	Effect of earthworm and biochar on microbial metagenomics	Fertile soil	Winding et al. (2016)
	Analytic comparison of metagenomes	Agricultural soil, Italy	Gigliucci et al. (2017)
	Assessing microbial response to change in redox potential in ASS using metagenomics	Acid sulphate soil	Su et al. (2017)
Transcriptomics	Identification of genes responsible for plant-microbe interactions using transcriptomic profiling	Two varieties of sugar beet, Boston, USA	Mark et al. (2005)
	First metatranscriptome from soil	South Western France, *Pinus pinaster* plantation	Bailly et al. (2007)
	First shotgun metatranscriptome from soil	An Rotball, Germany, lawn	Urich et al. (2008)
	Functional and structural analysis of a soil microbial community using transcriptomics approach	Tropical forest soil	Urich and Schleper (2011)
	Transcriptomic analysis of *Bacillus amyloliquefaciens* FZB42 in exposure to maize root exudates	Axenic hydroponic cultures	Fan et al. (2012)

(continued)

Table 3.1 (continued)

Tools	Analysis	Isolation source	References
	Analytical characterization of RNAs present in soybean libraries using metatranscriptomic approach	Field plantations and greenhouses	Molina et al. (2012)
	Functional transcripts constituted of transcripts that degrade plant constituents	Central France, Breuil-Chenue forests	Damon et al. (2012)
	Complex interaction of soil microbes with plants using transcriptome analysis	Semi-porous agricultural soil	Schenk et al. (2012)
	Interaction of *Pseudomonas aeruginosa* to a phosphate-deficient *Lolium perenne* rhizosphere: A transcriptional analysis	Manchester, UK	Zyśko et al. (2012)
	Transcriptome analysis of *Bacillus subtilis* OKB105 in exposure to rice seedlings	Axenic hydroponic cultures	Xie et al. (2015)
	Soil eukaryotic communities: A metatranscriptomic analysis	Grass soil sample	Yadav et al. (2016)
	Interpreting the relationship between bacterial species of *Methanothrix* and *Geobacter* using metatranscriptomic approach	Methanogenic rice paddy soils	Holmes et al. (2017)
	Identification of prokaryotic drivers of RNT in paddy soils using metatranscriptomic approach	Waterlogged paddy soils	Masuda et al. (2017)
	Analysing transcriptional responses of microbial communities on exposure to salt stress	Paddy soils	Peng et al. (2017)
Proteomics	Identification of eight enzymes involved in carbon cycling	Waldstein, Germany spruce forest	Schulze et al. (2005)
	Functional assessment and metaproteomic analysis of protein extracts from contaminated soil	Chlorinated soil	Benndorf et al. (2007)
	Application of proteomics to analyse the change in microbial communities in a hydrocarbon-polluted soil	Semiarid, diesel-contaminated soil	Bastida et al. (2010)
	Identification of 145 to 925 proteins using shotgun proteomics for amendment soils	Washington and California, laboratory study	Chourey et al. (2010)

(continued)

Table 3.1 (continued)

Tools	Analysis	Isolation source	References
	Effect of carbon amendment and methods of extraction on microbial proteins in soil	Quitman and Benfield soil	Taylor and Williams (2010)
	Identification of 16 proteins from toluene-amended soils, 8 from glucose-amended soils	Mississippi, agricultural soil	Williams et al. (2010)
	Comparative study on *Rehmannia glutinosa*-monocultured rhizosphere soil	Jiaozuo, Henan province, Central China	Wu et al. (2011)
	Effect of soil solid phases on the proteomic analysis of *Cupriavidus metallidurans*	Quartz sand, kaolinite, montmorillonite and artificial soil	Giagnoni et al. (2012)
	Analysis of ratoon sugarcane rhizospheric soil using metaproteomic approach	Ratoon sugarcane was regenerated from the germinating bud of the previous plant	Lin et al. (2013)
	Analysis of soils in semiarid environment using metaproteomic approach	Semiarid soil	Bastida et al. (2014)
	Metaproteomic analysis of Park Grass soil	Park Grass soil	Gerry et al. (2016)
	Interaction between bacterial functions within the rhizosphere of plants using metaproteomics	Serpentine soil	Mattarozzi et al. (2017)
	Application of functional metaproteomics in discovering bacterial biocatalysts	Fertile soil	Sukul et al. (2017)
	Quantification of biomass using metaproteomic approach	Alkaline soda soils	Kleiner et al. (2017)
Metabolomics	Detailed analysis of tobacco mosaic virus infection in *Nicotiana tabacum* leaves using metabolomic approach	Fertile soil	Choi et al. (2006)
	Analysis of methyl jasmonate-treated *Brassica rapa* leaves by two-dimensional NMR spectroscopy and metabolomics	Fertile agricultural soil	Liang et al. (2006)
	Analysis of the metabolomic response of *Brassica rapa* to preharvest bacterial contamination	Garden soil	Jahangir et al. (2008a)

(continued)

Table 3.1 (continued)

Tools	Analysis	Isolation source	References
	Assessment of metabolite accumulation in *Brassica rapa*	Agricultural soil	Jahangir et al. (2008b)
	Functional annotation of indigenous microbial genomes using carbon isotopomer-based metabolomics	Tropical and temperate forest soil	Fan et al. (2009)
	Metabolic response of tomato leaves on interaction with different pathogens	Garden soil	Gresa et al. (2010)
	Evaluation of the metabolic profile of tomato plants on long-term exposure to cadmium	Porous agricultural soil	Hédiji et al. (2010)
	Profiling of metabolites to record the responses of *Arabidopsis thaliana* to cadmium exposure	Artificial medium with variable Cd concentrations	Sun et al. (2010)
	Profiling of Arabidopsis seedlings in response to exogenous sinalbin and sulphur deficiency using metabolomic approach	Temperate soil	Zhang et al. (2011)
	Assessment of environmental pollution using metabolomic analysis of soil communities	Agricultural soil	Jones et al. (2013)
	Rhizospheric metabolomic analysis	Agricultural soil	Dam and Bouwmeester (2016)
	Analysis of fatty acid methyl esters in soil microbial communities using metabolomic approach	Temperate soil	Willers et al. (2016)
	Osmoprotection as strategy for drought tolerance in cowpea using metabolomics	Drought and well-watered soil	Goufo et al. (2017)
	Characterization of acid tolerance responses in *S. meliloti*: A metabolomic approach	Agricultural soil	Draghi et al. (2017)

Roche 454 genome sequencer which has a read capacity of 100 k and 1 M. In 2007, the most viable sequencers called Illumina sequencers were commercialized in the market as Solexa sequencers which use the principle of sequence by synthesis in combination with its amplification in a flow cell (Fedurco et al. 2006). On the other hand, whole genome sequencing of most of the organisms is not completely feasible due to technical difficulties and cost factors. Therefore, sequencing of the microbial mRNA has shredded immense light and importance especially on metatranscriptomics. Random sequencing of the mRNA samples is performed along with the required computational tools which help in the proper assessment of the microbial communities. Gene discovery and the assessment of the in situ activity of the microbes in the soil further strengthen the potential of transcriptomics (Warnecke and Hess 2009).

The entire characterization of the physiology and the activity of the microbial community cannot be determined by just metagenomics and metatranscriptomics. Detection of the soil samples needs to be done also at the protein level which is known as metaproteomics. Proteins are responsible for all the structural and functional relationships in an organism, and hence metaproteomics serves as an important tool in the profiling of the microbial proteins in the soil microorganisms. In soil microbial communities, there are various ways of determining the relation between a gene and its function such as genotyping, post-translational modifications, comparative and quantitative genomics, protein-protein interactions and protein cataloguing. To decontaminate the environmental samples contaminated with metallic and organic pollutants, proteomics-based assay is the primary approach.

Metabolites are small chemical molecules found in living organisms which directly assess its physiological nature by signal communication or energy consumption (Gieger et al. 2008). The only disadvantage may be the differentiating physical and chemical properties of the metabolites such as its molecular weight and size, solubility, extraction and detection. Hence, it is important to integrate the information from other Omics technologies with metabolomics for a more accurate and precise assessment of the soil microbiota.

A highly diversified heterogeneous microenvironment is created by the soil microbial community. Omics breakthroughs have revolutionized the detection and assessment of such microbial biota to a great extent. Currently, the use of Omics tools for soil microbial community is in its infant phase, but with a proper bioinformatic approach and specific functional characterization, it can mould the way to an exciting era of environmental microbiology.

3.2 Molecular Tools for Soil Community Analysis: Past, Present and Future Tools

3.2.1 Traditional Tools

Traditional approaches mostly relied upon cultivated clonal cultures, followed by sequencing of cloned specific genes (e.g. 16S rRNA gene) to produce a profile of diversity in the natural sample. Culture-based techniques when implemented for assessing the microbial diversity and the community dynamics of the soil samples sometimes act in an extremely biased manner by selecting a specific group of organisms. Such works revealed that vast majority of microbial diversity had been missed by culture-dependent methods as unculturable microorganisms compromise the majority of the planet's biodiversity. Unculturable microorganisms represent two out of three domains of life. In many environments, more than 99% of the microorganisms cannot be cultured by standard techniques (Great Plate Count Anomaly) which are recently called as the 'microbial dark matter'. Advancements in the culture-independent techniques have led to the evolution of a new era in microbial classification and analysis. Functional and structural information regarding microbial communities can be generated by direct isolation of nucleic acids, proteins, lipids and other relevant biomolecules from soil samples (Singh et al. 2009).

3.2.1.1 Microbial Analysis Using Partial Community Approaches

The basic principle of partial community approach is the PCR amplification of signature genes from the directly extracted DNA or RNA molecules from the soil sample which is followed by molecular fingerprinting (Hugenholtz 2002). Investigation and differentiation of different bacterial species can be done by amplifying 16S rRNA gene, 18S rRNA gene, recombinase A, gyrase beta subunit, heat-shock proteins, RNase beta subunit and other conserved genes (Ghebremedhin et al. 2008). The diversity and sensitivity of different soil types can also be determined using partial community approach in combination with different techniques such as genetic fingerprinting, clone library method and DNA microarrays (Fig. 3.2).

3.2.2 Cutting Edge Technologies: Microbial Analysis Using Whole Community Approaches

Compared to the molecular approaches which are PCR based, whole community microbial analysis gives a more transparent, comprehensive and clearer picture of the characterization, isolation, identification and genetic diversity of different taxonomic groups of phylogenetic microbial community. Sufficient resolution of 16S

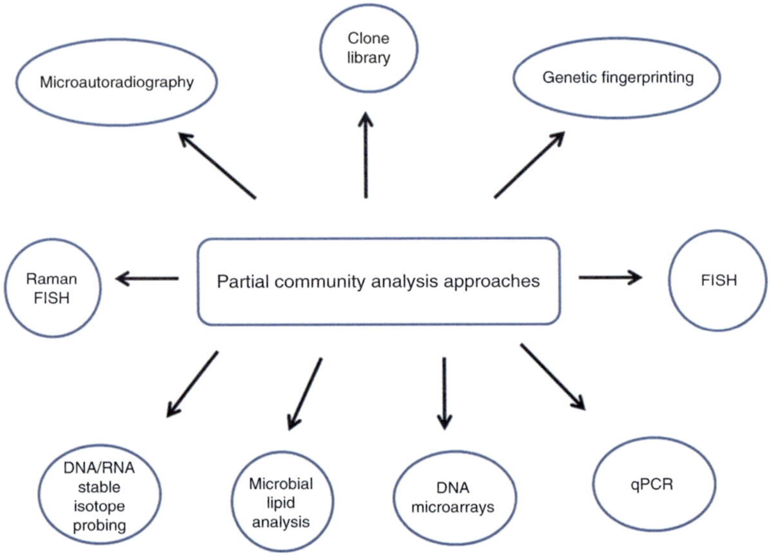

Fig. 3.2 Different molecular techniques used in partial community analysis approaches

rRNA gene is not obtained through partial community analysis of the microbial communities due to its conserved nature (Konstantinidis et al. 2006). Only a few or a single gene is targeted in partial community analysis, whereas in whole community analysis, a parallel massive sequencing technique is applied which assesses the potential function of the entire microbial community. There are various techniques to assess a microbial diversity using whole community analysis approach. These techniques may be used at a random or in different combinations to obtain a high throughput of the targeted microbial communities. Generally, the genetic information stored in the total extracted DNA from different environmental samples is analysed using the whole community approach (Fig. 3.3).

3.2.2.1 Postgenomic Approaches

Under in situ conditions, enough information about gene expression is not provided by the DNA-based molecular techniques (Wilmes and Bond 2006). Hence, new insights into the functional diversity of the microorganisms can be determined using metagenomic databases which also include sequences from the organisms that are uncultured. Therefore, a correlation between the functional and genetic similarities between different microbial communities can be determined using postgenomic approaches such as metaproteomics, proteogenomics and metatranscriptomics (Fig. 3.4).

Metagenomics

Metagenomics, community genomics or environmental genomics does not require the traditional knowledge about microbial communities but is the direct collection of

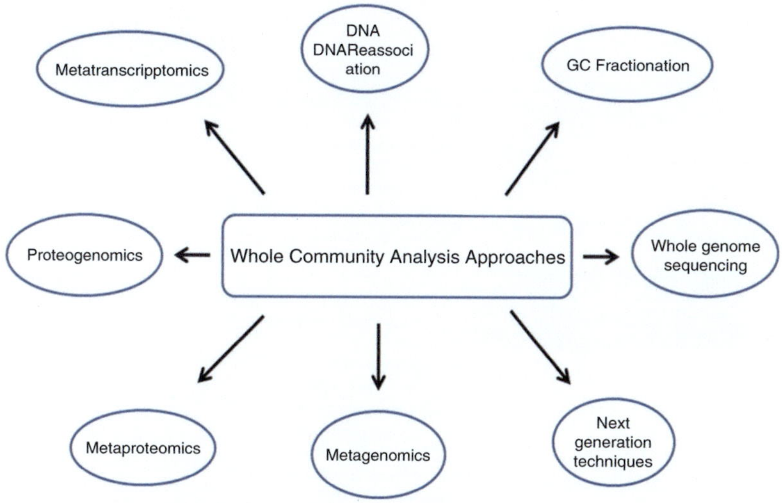

Fig. 3.3 Different molecular techniques used in whole community analysis approaches

microbial genomes from the environmental samples (Riesenfeld et al. 2004). Similar to the whole genome sequencing of a bacterium, metagenomics sequences the complete genome of the microbial communities isolated from the environment. Interaction of microorganisms with abiotic and biotic factors, uncultured organisms and their biochemical roles can hence be determined using metagenomics. Different functional molecules like antibiotics (terragine) and microbial enzymes (amylases, lipases, cellulases) can therefore be derived using functional metagenomic libraries (Rondon et al. 2000). Construction of a metagenome library is initiated by the isolation of DNA from the environmental sample, followed by using a vector for random cloning of the DNA fragments which is called as the shotgun cloning. Finally, screening of the positive clones is done after transforming the clones into a bacterial host. A major limitation in this technique is the frequent low-level expression of the active genes. Therefore, improved detection assays and high-throughput screening are necessary for a proper characterization of the microbial sample and its counterparts.

Recently, metagenome sequencing has been used to characterize microbial community structural as well as functional genes upon exposure to aerobic conditions in acid sulphate soils. Significant changes were observed both in parent material and topsoil upon incubating in aerobic conditions. Sulphur cycling genes abundantly increased in parent material, whereas there has been a significant decrease in archaea (Su et al. 2017). The relationship between taxonomic diversity and community composition still remains to be determined at the genetic level. Temperature, soil pH and other relevant environmental factors widely influence the structural properties of the microbial populations in the soil. Whole genome metagenomics can be used to assess the functional gene capacities of the bacterial communities.

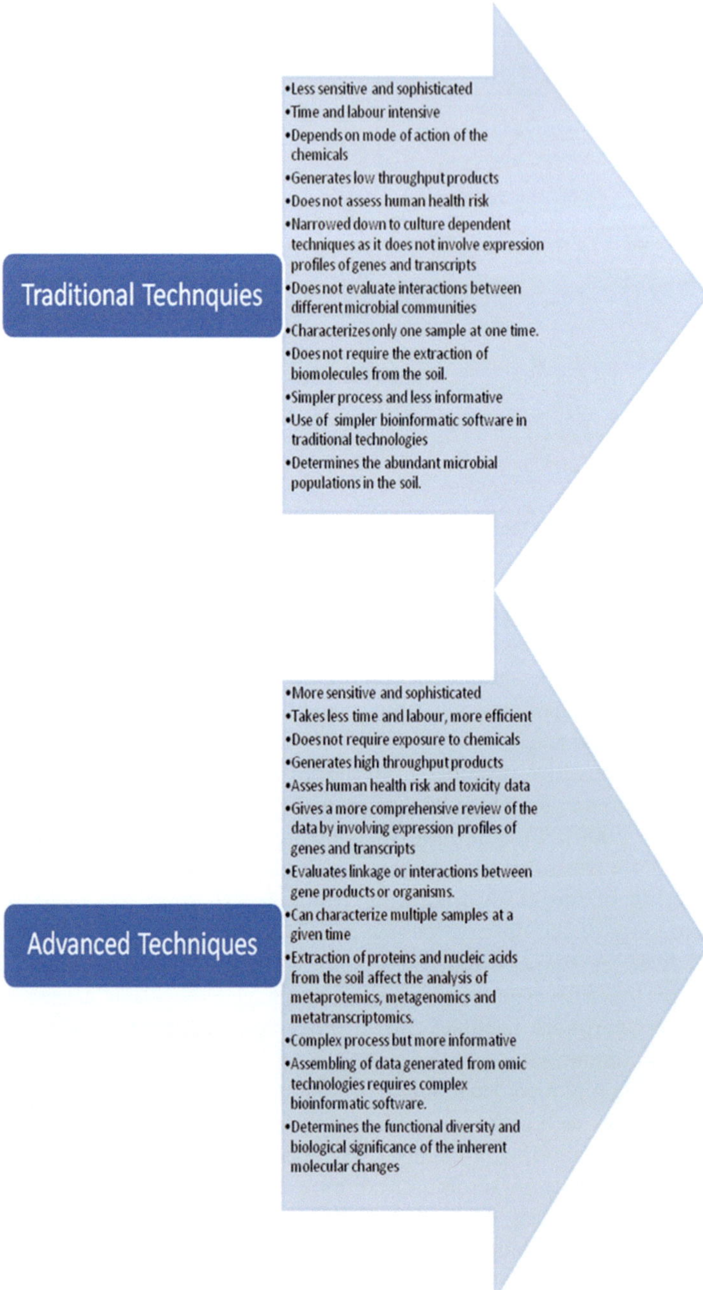

Fig. 3.4 Traditional vs modern techniques for the analysis of microbial diversity

Metatranscriptomics

Environmental transcriptomics or metatranscriptomics is the study of the changes in gene expression profile of microbes and their regulation in natural environments at a specific place and time by sequencing the mRNA transcripts which are randomly extracted from microbial communities (Moran 2009). A major setback to this technique is the absence of poly-A tails in prokaryotic mRNA transcripts which leads to the difficulty of obtaining the cDNA. Recent advances in metatranscriptomics have led to the development of a 'double RNA' method in which the abundantly rich RNA molecules, i.e. both mRNA and rRNA, are used to analyse the total phylogenetic pool of taxonomically and functionally relevant microbial community.

Therefore, recent advances in molecular tools have resulted in an impeccable enhancement in the field of microbial characterization and its inherent identification. A huge diversity has been noticed in the structural and functional characteristics of microorganisms. A major challenge in this field is the quantitative assessment of the vast number of microbial populations present in our ecosystem. Both the partial and whole community analysis approaches have their own advantages and limitations. Application of a combination of the required approaches may provide a better interrogation of the taxonomic diversities associated between different groups in the microbial world. Hence, an interdisciplinary approach should be followed between different Omics tools in order to provide an in-depth picture pertaining to environmental microbiology.

The possibility that *Geobacter* and *Methanothrix* species interact in methanogenic terrestrial environments through direct interspecies electron transfer was studied in rice paddy soils (Holmes et al. 2017). In relation to the global methane production, *Methanothrix* serves as the prominent microbial contributor, while very little is known about its ecology and physiology. In terrestrial ecosystems, the contribution made by *Methanothrix* to methane production may further extend beyond the conversion of methane from acetate. This serves the importance of DIET as a vital source of electrons for *Methanothrix* in rice paddy fields. Identification of reductive nitrogen transformation gene transcripts was done in waterlogged paddy soil through the help of metatranscriptomics (Masuda et al. 2017). Severe anoxic zones in waterlogged paddy soil aid the microbes to actively induce RNT or reductive nitrogen transformation. Temporal anaerobic conditions are the major characteristics of paddy soil by waterlogging and the occurrence of active anaerobic biogeochemical processes. The RNT-associated microbial diversity was investigated via RNA shotgun sequencing analysis without any prior PCR preparation. Seawater intrusion causes severe soil salinization along the coastal areas causing a worldwide threat to the cultivation of rice. Although there have been thorough studies on the detrimental impact on the rice yield and growth, little is known about the severity of the salinity affecting the function and structures of microbial paddy soil communities. Hence, controlled laboratory experiments were performed to examine their response to full-strength and half-strength seawater salinity using rice straw as a carbon source. Compared to control, the high and moderate salt stress suppressed the net consumption of methane and acetate production by 70% and 50%, respectively.

Among the family-level groups, significant changes were seen in the abundance of relative transcript corresponding to community-wide mRNA expression (Peng et al. 2017).

Eukaryotic organisms in the soil can express specific functions that can be studied by assessing the pool of specific eukaryotic polyadenylated mRNA directly isolated from environmental samples. Extraction of a desired quality of RNA from soil samples can be done using alternative extraction protocols. High-throughput sequencing can be used to convert the total amount of soil mRNA or RNA into cDNA. cDNA libraries can be constituted by cloning polyadenylated mRNA full-length cDNAs into plasmid expression vectors (Yadav et al. 2016). Hence, functional categories of gene can be screened using cDNA libraries. A combination of Illumina RNA sequencing of total RNA pyrosequenced SSU rRNA genes was also used to identify the microbial enzymes and populations associated with the degradation of the major components including exoskeletons of inhabiting peat arthropods and *Sphagnum*-derived litter comprising of xylan, chitin, pectin and cellulose (Ivanova et al. 2016). A fivefold or threefold increase in bacterial cell numbers was seen on addition of biopolymer to peat. An increased abundance of transcript genes was seen in xylan and pectin particularly enhancing central carbon and energy metabolism. On response to the availability of a particular polymer, the activity of the bacterial populations increased several folds. Thus, metatranscriptomics helps in analysing the diversity of microorganisms in the soil along with community composition that has largely remained concealed.

Metaproteomics

Study of proteins at a particular time on a large scale as expressed by the microbial community is done using environmental proteomics or metaproteomics (Wilmes and Bond 2006). Both quantitative and qualitative study of the diverse microbial species can be done using metaproteomics. Proteins give a more comprehensive view or a clearer picture compared to other biomolecules such as nucleic acids or lipids (Keller and Hettich 2009). Proteofingerprints are generated due to stress conditions which reflect the functional status of the microbial communities. For the extraction of abundant proteins, firstly the cells are separated from the microbial matrix by ultracentrifugation and then the cells undergo lysis. Community proteofingerprints are then generated by applying one-dimensional or two-dimensional electrophoresis. Finally, different analytical tools are applied to identify the protein spots. A quick identification of proteins is done by mass spectrometry in combination with liquid chromatography. Metaproteomics is very useful to assay the exact protein expression within soil community (Benndorf et al. 2007).

Recently, metaproteomics has been applied to analyse different environmental samples such as soil, sediments, freshwater and marine system. The various active metabolic processes have been studied through the help of metaproteomics which acts as a powerful tool in interpreting the relationships between the rigid processes (Siggins et al. 2012). The vast number of proteins expressed by different organisms in an ecosystem at a specific time is studied using this method. A strong functional link is been generated between the functional and the microbial phylogenetic community data by this (Bastida et al. 2014). On the other side, high diversity of microbial organisms, heterogeneity of the soil and improper metagenomic information negatively influence the above process (Wang et al. 2016). Complex carbohydrates, phenolic and lipid compounds and inorganic compounds such as clay and silt minerals along with humic acids possess a serious difficulty on the extraction of proteins from the soil (Keiblinger et al. 2016). Protein identification, depending on quality, capacity and design of the database of the chosen protein for analysis, acts as a major obstacle in the analysis of metaproteomics data. This can be overcome by using different methods of protein extraction and building a homogenous database to increase the characterization of proteins. Nowadays metaproteomics have the potential to reveal any function in an ecologically stable environment, hence acting as a stable process for environmental analysis.

A liquid chromatography, high-resolution mass spectrometry-based metaproteomic approach was used to interpret different metabolic functions of bacteria associated with plants acting as metal hyperaccumulator growing in serpentine soil which is naturally contaminated with chromium, cobalt and nickel along with rhizospheres of high metal-tolerating plants (Mattarozzi et al. 2017). Based on the genera recognized by 16S DNA profiling, an inherent bacterial protein database was built in particular. From LC-MS data, this was then further used for protein identification. Application of different extraction protocols to each soil sample generates variable information which on combination permits the identification of a vast number of proteins, corresponding to assigned functions in gene ontology categories. Proteins involved in the transportation of nutrients and metals and response to stimulus reveal a continuum of bacteria responsive to different microenvironment conditions. A key role in the transition to a post-petroleum bio-based economy lies with the bacterial biocatalysts. Actively screening for desired activities and analysing genetic information limit the capability of biocatalysts (Sukul et al. 2017). Hence, a combination of metagenomics and metaproteomics facilitates the direct and unmediated discovery of different biocatalysts in samples collected from the environment. Functional metaproteomics serves as an efficient method to analyse the biocatalytic activity from the proteome in an environmental sample. Also, quantifying biomass or cell numbers for individual populations helps in the proper detection of composition of a microbial community. Therefore, the application of abundance of proteins as a measure of biomass quantification for different populations serves as a method for assessing the composition of microbial populations. Methods for quantification of cell numbers are already available in the form of 16S rRNA gene amplicon sequencing, in situ hybridization and fluorescence. This method was used to analyse organisms from two different environments such as from saliva from different individuals and from microbial mats from two different alkaline Soda Lakes (Kleiner et al. 2017). Hence, metaproteomics serves as an important tool for analysis for interpreting the relationships between different organisms in an ecologically stable environment.

Proteogenomics

Proteogenomics is basically a combination of metaproteomics and metagenomic approaches (Banfield et al. 2005). The basic principle of this technique is the extraction of proteins and DNA from the same sample which allows for a greater understanding of the phylogenetic relationships between different microbial communities. Identification of inherent bacterial communities in the phyllosphere was done using proteogenomics (Delmotte et al. 2009).

3.3 Recent Advances: Integration of Genomics, Transcriptomics, Proteomics and Metabolomics for Studying Soil Microbial Community

3.3.1 Genomics in Soil Microbiology

The heterogeneous nature and the complex environment of the soil make it one of the most challenging environments to work with. An enormous amount of microorganisms is present in the soil, in variable combinations, most of them being uncultured till date due to its complex metabolic and genetic diversity. Since the early 1960s, there has been an incredible transformation in the approaches for characterization of the soil microbial community. Different bioprocessing techniques such as the measurement of flux of nitrogen and carbon and study of structure function relationships replaced the traditional methods of cell number counting of different organisms like bacteria, fungi and algae. Genomics, which focuses on the sequencing of the organisms' genome and analysing its characteristics, provided a new insight to the molecular approaches applied in the modern generation (Insam 2001).

It has been believed that the more functionally significant and numerically dominant species are the ones which have been isolated from the environment in pure cultures. This is generally a huge misconception because the ability to grow on artificial nutrient-rich media and to rapidly form colonies particularly in moderate temperatures and in aerobic conditions is the reason behind their isolation (Hugenholtz 2002). At first, some of the uncultivated lineages such as archaea have been detected using phylogenetic molecular surveys. High molecular weight, concentrated, clonable and purified DNAs have been extracted using a two-phase electrophoresis technique. The result suggested that the extracted archaea differ from its previously cultured counterparts. This paved the way for a more holistic and functional genomic approach (Quaiser et al. 2002).

Isolation of DNA from the soil along with the screening and production of libraries of clone forms the basis of soil metagenomics. Detection, assessment and cultivation of large pools of the diverse genetic reservoir among soil microbial community have already left a noticeable impact on the discovery of novel genetic biomolecules. Still, more than half of the organisms in the entire ecosystem remains uncultured. The 16s rRNA genes revolutionized the complete system by characterizing the physiology and biochemistry besides the genetic makeup of the soil microorganisms. The first comprehensive metagenome of soil was done by Tringe et al. in 2005. Comparison of genes was done between the soil obtained from Minnesota and other environments (Myrold et al. 2013). Nowadays shotgun sequencing methods are being used which directly sequences the DNA extracted and does not require the formation of libraries. When combined with Illumina sequencing systems, it produces a saturated genome size of round about 4.0 Gbp and many more such sequences comparatively. Both assembly contigs and individual reads contribute to the screening of functional potential of the metabolic activity of the soil microbiota. It further helps in differentiating between microorganisms in both complex and heterogeneous environments. Different soil metagenome sequencing is given in Table 3.1.

Till date, about 48 soil metagenomes are registered online where not all of them are shotgun metagenomes and most of them are still not published. In 2012, from the submerged sediments of four mangrove forests in Brazil, metagenomes were generated which constructed carbon, nitrogen and sulphur metabolic pathways (Andreote et al. 2012). Recently, to infer the interaction of networks between microbial communities, random matrix theory has been developed which interprets the change in the network in correlation with the elevated levels of carbon dioxide (Zhou et al. 2011). Also, the presence of different microbial phyla is determined in relation with their strategies of life history. Therefore, the use of trait-based models of biogeochemical cycles in soil to incorporate metagenomic data may open a wider road for metagenome interpretation (Allison 2012).

The ultimate aim of metagenomic studies is linking of the composition of the microbial community to its function. Proper assessment of the heterogeneous soil environment is directly related to the microbial activities associated with it. Hence, integration of other Omics tools along with the DNA-RNA-protein continuum approach paves the way for a more enumerated and advanced technique of soil microbial characterization and assessment.

3.3.2 Transcriptomics in Soil Microbiology

The central dogma theory states the transcription of DNA into RNA and the translation of RNA into proteins. The microbial activity of the microorganisms in soil is generally a result of their gene expression, i.e. transcription of DNA into RNA, which is also responsible for the change in their physiological behaviour in accordance to different environmental variation. The relative activity of the microbes can be measured by the ratio of their rRNA/DNA as the amount of rRNA in prokaryotic organisms is relatively constant. The growth rate of cellular microorganisms is also directly proportional to their cellular rRNA (Blazewicz et al. 2013). It is also seen that the composition of microbes in a heterogeneous environment, like soil, differs on the basis of their genetic material, i.e. DNA or RNA. Several studies suggest that such differences have also been observed between different transcripts and genes when metatranscriptomes and metagenomes have been obtained from soil or other marine environments (Gilbert et al. 2008) (Table 3.1).

Currently, there is not much advancement or publication of studies related to soil metatranscriptomics. The isolation of poly-A tail of eukaryotic mRNA acts as an advantage for sequencing of the genetic material. RNA-seq or shotgun metatranscriptomics have been used in a few studies (Croucher and Thomson 2010). For a significant analysis of mRNA sequences, hybridization of both enrichment and non-enrichment process has been adopted (Urich et al. 2008; Stewart et al. 2011). In-depth exploration needs to be done to provide insights regarding the function of cellular microbes in other microbial systems which are far more complex in nature. The first metatranscriptome for soil was developed in South Western France using soil from *Pinus pinaster* plantation on coastal sand dune (Bailly et al. 2007) (Table 3.1).

The functioning of the terrestrial ecosystem is directly related to the soil microbial activity. Hence, at the time of sampling, the transcription profile of a discrete population of microbial biota presents us with a snapshot of the metatranscriptomic studies. Difficulty in generating high-quality cDNA from the extracted RNA may act as a limitation for the utilization of metatranscriptomics. Therefore, the potential of metatranscriptomes lies in the adoptability of the entire microbial community to different environmental conditions, and it further inoculates the source for different genes of biotechnological interest.

3.3.3 Proteomics in Soil Microbiology

Proteins mediate most of the important functions in a microbial community. Hence, metaproteomics has become the utmost valuable and direct Omics tool for measuring the microbial activity in an environment mediated by both biotic and abiotic factors. For the extraction of proteins from the soil, a more evolved and a steady methodology have been developed besides the use of metagenomic and metatranscriptomics (Bastida et al. 2009). The strong interaction of proteins with different minerals present in the soil and with other biomolecules can result as a major limitation of the protein extraction methods. The specifications of soil may be frequently required as a standard method for the extraction of proteins from the soil which has still not yet been developed.

Based on gel isolation methods, till date only a few proteins have been extracted from soil to determine the soil microbial activity (Table 3.1). The combination of a more advanced protein separation method called as the two-dimensional liquid chromatography along with tandem mass spectrometry forms the basis of a more sophisticated form of metaproteomics called as the shotgun metaproteomics which generates proteins from hundreds to thousands in numbers (Chourey et al. 2010). Compared to metagenomics and metatranscriptomics, very few proteins have been identified till date which are generally abundant in nature or are associated with functional activities such as housekeeping apart from the ones which are associated with biogeochemical processes and have a specific enzyme activity.

Four proteins were identified in the laboratory study of compost soils in Leipzig, Germany, using NaOH and phenol extraction methods followed by their analysis using SDS and LC chromatography (Benndorf et al. 2007). Similarly, eight enzymes associated with carbon cycling were analysed using the same method where the extraction of the proteins was done by dissolving the minerals of the soil in 10% HF (Schulze et al. 2005). A laboratory study was done in Washington and California using soils with or without addition of known bacterial species. In this method, the extraction of proteins was done by SDS and TCA followed by analysis using two-dimensional LC-MS. About 145–925 proteins were identified using shotgun proteomics for amended soils (Chourey et al. 2010). Eight proteins from glucose-amended soils and 16 proteins from toluene-amended soils were first identified in Mississippi using indirect phenol extraction method and SDS-PAGE followed by MALDI/TOF (Williams et al. 2010).

High yield and proper resolution of proteins is needed for the appropriate analysis of the soil samples using metaproteomics which is an ambitious task. For a high-throughput resolution, the extraction and preparation of protein samples is a crucial step. Application of metaproteomics to soil faces several challenges due to its heterogeneous nature, complexity, spatial distribution of protein, dynamic nature of the microbial community, abundancy of known and unknown proteins in the soil, adherence of extracellular enzyme to minerals in the soil or the entrapment of colloids which are generally humic in nature (Nannipieri 2006). Recent advances in soil metaproteomics include extraction of proteins by indirect method in which the extraction of microbial cells is done from the sol matrix and prior to the protein analysis; the cells undergo enrichment (Taylor and Williams 2010). Minimal changes in the composition of the proteome during the preparation of soil sample and completeness of the extracted protein are the reasons behind the success of direct extraction of proteins compared to the indirect methods (Benndorf et al. 2007). Therefore, more efficient and reliable techniques for protein extraction, evaluation and metaproteome analysis need to be developed which can accurately determine the cellular activity of the microbial community in spite of the complex and the heterogeneous nature of both the proteins and the soil.

3.3.4 Metabolomics in Soil Microbiology

Metabolomics characterizes soil microbial activity and its interaction with its immediate environment and abiotic pressures and also interprets the response of a specific organism to other biological factors. Population-based or individual studies are done to determine the adaptability of different groups of phylogenetically diverse microbial communities. Metabolites are low molecular weight biomolecules and generally organic in nature and occur naturally in a biofluid, a cell or a tissue. The functional status of an organism can be reported by the measurements done by metabolomics. A metabolome or a set of metabolites produced on interaction with an environmental stimulus is the basis for the comprehensive study of metabolomics (Miller 2007). An indication of an extremely sensitive external stress can be determined by the change in metabolome which is due to the changes in environmental stimuli (Ankley et al. 2006).

Nowadays, assessment of ecological risks can be done using metabolomics. A toxicant, its mode of action and the potential threats associated with it can be studied by the implementation of metabolomics. Exposure to different inorganic and organic pollutants leads to the development of different metabolic profiles among cellular organisms (Table 3.1). Thus, PCA or the principal component analysis is used to detect such differences among organisms (Jones et al. 2014). Assessment of soil contamination using metabolomics is still in its infant stage (Viant 2009). Microbial toxicity increases by the uptake of heavy metal ions. This changes the biochemical as well as the physiological nature of metabolites depending on the chemical nature of the adsorbed pollutant. Metabolomics provides an unbiased assessment of the soil microbial activity in a given specific condition (Rochfort 2005). In the biogeochemical cycle of nutrients, metabolic breakdown and transport of metabolites occurs in and out of the cell which enhances the assessment of the activity of different microbial communities.

About 240 projects on metagenomics have been completed till July 2011 as stated by DOE Joint Genome Institute. To study the dynamics of intrinsic bioremediation in different environments, a system-wide approach is doing rounds utilizing the basis of metabolomics. Metagenomics and metatranscriptomics are both complemented by metabolomics. Biochemical pathways have already undergone a wide change in their dimensions owing to the vast applications of metabolomics. The complication of modelling and reconstruction of microbial metabolism involve expertise evaluations and a lot of repetitive steps. Therefore, proper knowledge about microbial mechanisms and variant nature of metabolites is necessary for an accurate, reliable and precise determination of soil microbial activity.

At molecular level, metabolomics offers additional advantage for studying interactions between environment and organisms along with their assessment of health and functionality. An organism's inherent functional status can be generated from the data of metabolomics measurements. It can be further related to higher folds of biological complexity, i.e. practically to organism's phenotype. Discovering relations between least expected metabolic responses and relationships can itself lead to the generation of a hypothesis. Metabolomics has numerous applications in environmental ecology and molecular sciences, ranging from understanding the response of organisms to abiotic stressors including both anthropogenic factors such as pollution control and natural factors such as to study biotic-biotic interactions, temperature, herbivory and infections.

Particularly in the field of environmental sciences, metabolomics have several major applications including disease monitoring and diagnosis, assessment of risk on exposure to toxicants, metabolic responses to different levels of environmental stress and in the development of biomarkers. This approach has also been used to study the challenges faced during measurement of metabolic variability and metabolites along with proper interpretation of complex multivariate metagenomics data in aquatic organisms (Viant 2009). Simultaneous measurement of multiple complex metabolites can be done during metabolomics studies using mass spectrometry and NMR spectrometry which are inherently conclusive analytical techniques of measurement. This is followed by a precise statistical analysis using repeated univariate or multivariate tests. Several changes were identified in different metabolic pathways of acid-challenged *S. meliloti.* Correspondingly other bacteria undergoing a range of abiotic stresses could be related to the same. Biosynthesis of exopolysaccharides, modifications in the pentose phosphate pathway and intermediates in the myo-inositol degradation are some of the observable modification changes. In rhizobia, as a part of metabolic adaptation, these modifications are associated with improved competitiveness of nodulation and enhanced acid-tolerant phenotypes (Draghi et al. 2017).

Osmoprotection and osmotic adjustment was studied in model species of cowpea using metaproteomics (Goufo et al. 2017). Different cultivars were grown in drought conditions, watered soil and watering post-drought conditions via fruit formation. Drought tolerance in plants is usually done through the production of organic solutes which can either act as radical scavengers for protecting the metabolic functions or maintain an optimum turgor by acting as compatible osmolytes. Out of the 88 metabolites investigated, the strongest response to drought conditions was shown by quercetin, galactinol and proline as already highlighted by analysis done by multivariate and their relation with yield generating beneficial and positive effects. To cope with drought in the aerial parts of the plant, these served as a very conservative strategy as they accumulated in a similar pattern in the leaves but differently in the roots. Changes in protective mechanisms reflected the changes in energy investment with the accumulation of metabolites. Also by analysing the difference in the metabolic profile of plants, salt-tolerant mechanisms can be studied which can be a fruitful tool for sustaining enhanced crop production worldwide. Generally, salinity acts a major environmental stress limiting the growth of plants and reducing their yield. Therefore, these findings provide an important foundation for the utilization, development and protection of the natural plant resources.

3.4 Challenges in 'Omics' Approaches in Soil Microbiology Analysis

Application of 'Omics' in soil microbiology is limited by several distinguished factors. Heterogeneity of soil samples creates the most difficult problem in getting representative soil sample for extraction and purification of nucleic acid or other relevant molecules for structural and functional microbial analysis. Due to the presence of adequate nutrients and other physicochemical environments in the specific 'hotspots' or 'microniches', microbes prefer this microenvironment (Nannipieri 2006). This spatial heterogeneity makes the preference of study on special types of microorganisms and their interactions with microhabitats (Fierer et al. 2005). To overcome the special heterogeneity of soil, researchers have developed the microsampling and average across microhabitats strategies (Wellington et al. 2003). Temporal variability (i.e. seasonal and diurnal variation in temperature, humidity and other factors) further increased the complexity of metagenome or metaproteome or metatranscriptome study. Another major problem in the extraction processes of nucleic acids and biomolecules is the inefficient cell lysis which not only affects the yield but also the completeness of soil microbial communities (Delmont et al. 2011).

In addition, the organic compounds like humic acid present in soil often inhibit the extraction as well as PCR process. Omics analysis is also to some extent limited to the PCR biasness. The small read lengths generated by the present sequencing systems actually produce a series of noncontiguous genetic fragments creating a problem during assembling them. Handling large amounts of data or reads generated by the advanced sequencing techniques really needs expertise in bioinformatics and high-end computer system (Vinaixa et al. 2016). Fruitful application of Omics data in soil microbiology needs thorough expertise in bioinformatics as well as soil microbiology to interpret their actual interactions. Also, appropriate modelling and statistical tools are required for correct interpretation of the vast amount of data generated through Omics methods. Application of different network analysis in recent times has shown a promisable path in this genre (Zhou et al. 2011). A major challenge for bioinformaticians is the successful grappling of the generated complex datasets. The basis of the microbial processes in soil can be understood by studying the interactions between the microbes or the microbes and their abiotic environment. The synchronicity and the timescales for the responses of different metagenomes need to be revised so as to gain a proper understanding about microbial metabolisms. Hence, a meta-omic approach has the capability to capture the functional and taxonomical diversity and assess the interactions between metabolic processes of different taxa of soil microbes.

3.5 Future Prospects of 'Omics' in Soil Microbiology

With the development and application of molecular genomic tools, the field of microbial ecology is undergoing unprecedented changes. Postgenomic molecular approaches enable us to interrogate the structural and functional diversity of soil microbial communities and reveal that we have only scratched the surface of the genetic and metabolic diversity present in the most abundant organisms of the Earth, the *prokaryotes*. Several important questions such as 'How many microbial species are there on the Earth?', 'What is the extent of metabolic diversity in natural microbial communities?' and 'How microbial communities are governed by biological, chemical and physical factors?' remain to be understood. Understanding the functional roles of uncultured organisms still remains a daunting task, as most of the genes identified have no homologous representatives in databases. Although considerable progress has been made in the characterization of microbial communities by the application of metagenomic, metatranscriptomic and proteogenomic approaches, many technical challenges remain. An interdisciplinary system approach embracing several 'Omics' technologies to reveal the interactions between genes, proteins and environmental factors will be needed to provide new insights into soil microbiology. Development of multi- 'Omics' approaches will be a high-priority area of research in the coming years. The field of soil microbial ecology has been revolutionized by the combination of all the above Omics technologies. These new approaches allow linkage between microbial diversity and their specific ecological interactions in the environment which strengthens our in-depth understanding of the subject. This chapter presents an overview of the recent advancement in Omics technologies and its challenges and future prospects in understanding the soil microbial ecology.

References

Allison SD (2012) A trait-based approach for modelling microbial litter decomposition. Ecol Lett 15:1058–1070

Altschul SF, Gish W, Miller W, Myers EW, Lipman DJ (1990) Basic local alignment search tool. J Mol Biol 215:403–410

Amellal N, Burtin G, Bartoli F, Heulin T (1998) Colonization of wheat roots by an exopolysaccharide-producing Pantoea agglomerans strain and its effect on rhizosphere soil aggregation. Appl Environ Microbiol 64:3740–3747

Andreote FD, Jimenez DJ, Chaves D, Dias ACF, Luvizotto DM, Andreote FD, Fasanella CC, Lopez MV, Baena S, Taketani RG, Melo ISD (2012) The microbiome of Brazilian mangrove sediments as revealed by metagenomics. PLoS One 7(6):e38600. https://doi.org/10.1371/journal.pone.0038600.

Ankley GT, Daston GP, Degitz SJ, Denslow ND, Hoke RA, Kennedy SW, Miracle AL, Perkins EJ, Snape J, Tillitt DE, Tyler CR, Versteeg D (2006) Toxi-cogenomics in regulatory ecotoxicology. Environ Sci Technol 40:4055–4065

Bailly J, Tachet LF, Verner MC, Debaud JC, Lemaire M, Louvel MW, Marmeisse R (2007) Soil eukaryotic functional diversity, a metatranscriptomic approach. ISME J 1:632–642

Banfield DJF, Verberkmoes NC, Hettich RL, Thelen MP (2005) Proteogenomic approaches for the molecular characterization of natural microbial communities. OMICS 9:301–333

Bastida F, Moreno JL, Nicolas C, Hernandez T, Garcia C (2009) Soil metaproteomics: a review of an emerging environmental science. Significance, methodology and perspectives. Eur J Soil Sci 60:845–885

Bastida F, Nicolas C, Moreno JL, Hernandez T, Garcia C (2010) Tracing changes in the microbial community of a hydrocarbon-polluted soil by culture-dependent proteomics. Pedosphere 20:479–485

Bastida F, Hernández T, García C (2014) Metaproteomics of soils from semiarid environment: functional and phylogenetic information obtained with different protein extraction methods. J Proteome 101:31–42

Benndorf D, Balcke GU, Harms H, Bergen MV (2007) Functional metaproteome analysis of protein extracts from contaminated soil and groundwater. ISME J 1:224–234

Blazewicz SJ, Barnard RL, Daly RA, Firestone MK (2013) Evaluating rRNA as an indicator of microbial activity in environmental communities: limitations and uses. ISME J 7 (11):2061–2068. https://doi.org/10.1038/ismej.2013.102

Chen LS, Figueredo A, Pedrosa FO, Hungria M (2000) Genetic characterization of soybean rhizobia in Paraguay. Appl Environ Microbiol 66:5099–5103

Choi YH, Kim HK, Linthorst HJM, Hollander JG, Lefeber AWM, Erkelens C, Nuzillard JM, Verpoorte R (2006) NMR Metabolomics to revisit the tobacco mosaic virus infection in Nicotiana tabacum leaves. J Nat Prod 69:742–748

Chourey K, Jansson J, VerBerkmoes N, Shah M, Chavarria KL, Tom LM, Brodie EL, Hettich RL (2010) Direct cellular lysis/protein extraction protocol for soil metaproteomics. Proteome Res 9:6615–6622

Chung EJ, Lim HK, Kim JC, Choi GJ, Park EJ, Lee MH, Chung YR, Lee SW (2008) Forest soil metagenome gene cluster involved in antifungal activity expression in Escherichia coli. Appl Environ Microbiol 74:723–730

Cieśliński H, Białkowskaa A, Tkaczuk K, Długołecka A, Kur J, Turkiewicz M (2009) Identification and molecular modeling of a novel lipase from an Antarctic soil metagenomic library. Pol J Microbiol 58:199–204

Croucher NJ, Thomson NR (2010) Studying bacterial transcriptomes using RNA-seq. Curr Opin Microbiol 13:619–624

Dam NM, Bouwmeester HJ (2016) Metabolomics in the Rhizosphere: tapping into belowground chemical communication. Trends Plant Sci 21:256–265

Damon C, Lehembre F, Desfeux CO, Luis P, Ranger J, Tachet LF, Marmeisse R (2012) Metatranscriptomics reveals the diversity of genes expressed by eukaryotes in forest soils. PLoS One 7(1):e28967. https://doi.org/10.1371/journal.pone.0028967

Delmont TO, Robe P, Cecillon S, Clark IM, Constancias F, Simonet P, Hirsch PR, Vogel TM (2011) Accessing the soil metagenome for studies of microbial diversity. Appl Environ Microbiol 77:1315–1324

Delmotte N, Knief C, Chaffron S, Innerebner G, Roschitzki B, Schlapbach R, Mering CV, Vorholt JA (2009) Community proteogenomics reveals insights into the physiology of phyllosphere bacteria. Proc Natl Acad Sci U S A 106:16428–16433

Draghi WO, Papa MFD, Barsch A, Albicoro FJ, Lozano MJ, Pühler A, Niehaus K, Lagares A (2017) A metabolomic approach to characterize the acid-tolerance response in *Sinorhizobium meliloti*. Metabolomics 24:13–71

Fan TWM, Bird JA, Brodie EL, Lane AN (2009) 13C Isotoper – based metabolomics of microbial groups isolated from two forest soils. Metabolomics 5:108–122

Fan B, Carvalhais LC, Becker A, Fedoseyenko D, Wirén NV, Borriss R (2012) Transcriptomic profiling of Bacillus amyloliquefaciens FZB42 in response to maize root exudates. BMC Microbiol. https://doi.org/10.1186/1471-2180-12-116

Fedurco M, Romieu A, Williams S, Lawrence I, Turcatti G (2006) BTA, a novel reagent for DNA attachment on glass and efficient generation of solid-phase amplified DNA colonies. Nucleic Acids Res 34(3):e22. https://doi.org/10.1093/nar/qnj023

Feng YY (2013) Omics breakthroughs for environmental microbiology. Omics Environ Microbiol 40:18–33

Fierer N, Jackson JA, Vilgalys R, Jackson RB (2005) Assessment of soil microbial community structure by use of taxon-specific quantitative PCR assays. Appl Environ Microbiol 71:4117–4120

Fierer N, Breitbart M, Nulton J, Salamon P, Lozupone C, Jones R, Robeson M, Edwards RA, Felts B, Rayhawk S, Knight R, Rohwer F, Jackson RB (2007) Metagenomic and small-subunit rRNA analyses reveal the genetic diversity of bacteria, archaea, fungi, and viruses in soil. Appl Environ Microbiol 73:7059–7066

Garbeva P, Veen JAV, Elsas JDV (2004) Microbial diversity in soil: selection microbial populations by plant and soil type and implications for disease suppressiveness. Annu Rev Phytopathol 42:243–270

Gerry Q, Ed D, Stefan D, Peter M, Ingrid H, Richard W, Rhys A, Tom D, Lewis F, Andrea GS, Geertje VK (2016) The metaproteome of "Park Grass" soil – a reference for EU soil science. Copernicus 82:15981–15982

Ghebremedhin B, Layer F, König W, König B (2008) Genetic classification and distinguishing of Staphylococcus species based on different partial gap, 16 rRNA, hsp60, rpoB, sodA, and tuf gene sequences. J Clin Microbiol 46:1019–1025

Giagnoni L, Magherini F, Landi L, Taghavi S, Lelie DVD, Puglia M, Bianchi L, Bini L, Nannipieri P, Renella G, Modesti A (2012) Soil solid phases effects on the proteomic analysis of Cupriavidus metallidurans CH34. Biol Fertil Soils 48:425–433

Gieger C, Geistlinger L, Altmaier E, Angelis MHD, Kronenberg F, Meitinger T, Mewes HW, Wichmann HE, Weinberger KM, Adamski J, Illig T, Suhre K (2008) Genetics meets metabolomics: a genome-wide association study of metabolite profiles in human serum. PLoS Genet 4(11):e1000282. https://doi.org/10.1371/journal.pgen.1000282

Gigliucci F, Brambilla G, Tozzoli R, Michelacci V, Morabito S (2017) Comparative analysis of metagenomes of Italian top soil improvers. Environ Res 155:108–115

Gilbert JA, Field D, Huang Y, Edwards R, Li W, Gilna P, Joint I (2008) Detection of large numbers of novel sequences in the metatranscriptomes of complex marine microbial communities. PLoS One 3(8):E3042. https://doi.org/10.1371/journal.pone.0003042

Goufo P, Pereira JMM, Jorge TF, Correia CM, Oliveira MR, Rosa EAS, Antonio C, Trindade H (2017) Cowpea (Vigna unguiculata L. Walp.) Metabolomics: osmoprotection as a physiological strategy for drought stress resistance and improved yield. Front Plant Sci. https://doi.org/10.3389/fpls.2017.00586

Gresa MPL, Maltese F, Bellés JM, Conejero V, Kim HK, Choi YH, Verpoorte R (2010) Metabolic response of tomato leaves upon different plant–pathogen interactions. Phytochem Anal 21:89–94

Hassink J, Bouwman LA, Zwart KB, Bloem J, Brussaard L (1993) Relationships between soil texture, physical protection of organic matter, soil biota, and C and N mineralization in grassland soils. Geoderma 57:105–128

Hédiji H, Diebali W, Cabasson C, Maucourt M, Baldet P, Bertrand A, Zoghlami LB, Deborde C, Moing A, Brouquisse R, Chaibi W, Gallusci P (2010) Effects of long-term cadmium exposure on growth and metabolomic profile of tomato plants. Ecotoxicol Environ Saf 73:1965–1974

Holmes DE, Shrestha PM, Walker DJF, Dang Y, Nevin KP, Woodard TL, Lovley DR (2017) Metatranscriptomic evidence for direct interspecies electron transfer between *Geobacter* and *Methanothrix* species in methanogenic rice paddy soils. Appl Environ Microbiol 83:9–17

Howe AC, Jansson JK, Malfatti SA, Tringe SG, Tiedje JM, Brown CT (2014) Tackling soil diversity with the assembly of large, complex metagenomes. PNAS 111:4904–4909

Hugenholtz P (2002) Exploring prokaryotic diversity in the genomic era. Genome Biol 3(2). https://doi.org/10.1186/gb-2002-3-2-reviews0003

Insam H (2001) Developments in soil microbiology since the mid 1960s. Geoderma 100:389–482

Ivanova AA, Wegner CE, Kim Y, Liesack W, Dedysh SN (2016) Identification of microbial populations driving biopolymer degradation in acidic peatlands by metatranscriptomic analysis. Mol Ecol 25:4818–4835

Jahangir M, Farid IBA, Choi YH, Verpoorte R (2008a) Metal ion-inducing metabolite accumulation in Brassica rapa. J Plant Physiol 165:1429–1437

Jahangir M, Kim HK, Choi YH, Verpoorte R (2008b) Metabolomic response of Brassica rapa submitted to pre-harvest bacterial contamination. Food Chem 107:362–368

Jones OAH, Sdepanian S, Lofts S, Svendsen C, Spurgeon DJ, Maguire ML, Griffin JL (2013) Metabolomic analysis of soil communities can be used for pollution assessment. Environ Toxicol Chem 33:61–64

Jones OAH, Sdepanian S, Lofts S, Svendsen C, Spurgeon DJ, Maguire ML, Griffin JL (2014) Metabolomic analysis of soil communities can be used for pollution assessment. Environ Toxicol Chem 33:61–64

Júnior GVL, Noronha MF, Sousa STP, Cabral L, Domingos DF, Saber ML, Melo IS, Oliveira VM, Baldrian P (2017) Potential of semiarid soil from Caatinga biome as a novel source for mining lignocellulose-degrading enzymes. FEMS Microbiology 93:62–68

Keiblinger KM, Fuchs S, Boltenstern SZ, Riedel K (2016) Soil and leaf litter metaproteomics – a brief guideline from sampling to understanding. FEMS Microbiol Ecol 92:66–74

Keller M, Hettich R (2009) Environmental proteomics: a paradigm shifts in characterizing microbial activities at the molecular level. Microbiol Mol Biol Rev 73:62–70

Kim JS, Lim HK, Lee MH, Park JH, Hwang EC, Moon BJ, Lee SW (2009) Production of porphyrin intermediates in Escherichia coli carrying soil metagenomic genes. FEMS Microbiol Lett 295:42–49

Kleiner M, Thorson E, Sharp CE, Dong X, Liu D, Li C, Strous M (2017) Assessing species biomass contributions in microbial communities via metaproteomics. bior Xiv. https://doi.org/10.1101/130575

Konstantinidis KT, Ramette A, Tiedje JM (2006) The bacterial species definition in the genomic era. Philos Trans R Soc B 361:1929–1940

Liang YS, Choi YH, Kim HK, Linthorst HJM, Verpoorte R (2006) Metabolomic analysis of methyl jasmonate treated Brassica rapa leaves by 2-dimensional NMR spectroscopy. Phytochemistry 67:2503–2511

Lim HK, Chung EJ, Kim JC, Choi GJ, Jang KS, Chung YR, Cho KY, Lee SW (2005) Characterization of a forest soil metagenome clone that confers indirubin and Indigo production on Escherichia coli. Appl Environ Microbiol 71:7768–7777

Lin W, WuL LS, Zhang A, Zhou M, Lin R, Wang H, Chen J, Zhang Z, Lin R (2013) Metaproteomic analysis of ratoon sugarcane rhizospheric soil. BMC Microbiol. https://doi.org/10.1186/1471-2180-13-135

Mark GL, Dow JM, Kiely PD, Higgins H, Haynes J, Baysse C, Abbas A, Foley T, Franks A, Morrissey J, Gara F (2005) Transcriptome profiling of bacterial responses to root exudates identifies genes involved in microbe-plant interactions. G. Louise Mark 102:17454–17459

Masuda Y, Itoh H, Shiratori Y, Isobe K, Otsuka S, Senoo K (2017) Predominant but previously looked prokaryotic drivers of reductive nitrogen transformation in paddy soils, revealed by metatranscriptomics. Microbes Environ 62:1347–4405

Mattarozzi M, Manfredi M, Montanini B, Gosetti F, Sanangelantoni AM, Marengo E, Careri M, Visioli G (2017) A metaproteomic approach dissecting major bacterial functions in the rhizosphere of plants living in serpentine soil. Anal Bioanal Chem 409:2237–2339

Miller MG (2007) Environmental metabolomics: SWOT analysis (strengths, weaknesses opportunities and threats). J Proteome Res 6:540–545

Mocalli S, Benedetti A (2010) Exploring research frontiers in microbiology: the challenge of metagenomics in soil microbiology. Res Microbiol 161:497–505

Molina LG, Fonseca GCD, Morais GLD, Oliveira LFVD, Carvalho JBD, Kulcheski FR, Margis R (2012) Metatranscriptomic analysis of small RNAs present in soybean deep sequencing libraries. Genet Mol Biol 35:292–303

Moran MA (2009) Metatranscriptomics: eavesdropping on complex microbial communities. Microbe 4:329–335

Myrold DD, Zeglin LH, Jansson JK (2013) The potential of metagenomic approaches for understanding soil microbial processes. SSSAJ 78:3–10

Nacke H, Will C, Herzog S, Nowka B, Engelhaupt M, Daniel R (2011) Identification of novel lipolytic genes and gene families by screening of metagenomic libraries derived from soil samples of the German Biodiversity Exploratories. FEMS Microbiol Ecol 78:188–201

Nannipieri P (2006) Roles of stabilised enzyme in microbial ecology and enzyme extraction from soil and potential applications in soil proteomics. Soil Biol 8:75–94

Peng J, Wegner CE, Liesack W (2017) Short-term exposure of paddy soil microbial communities to salt stress triggers different transcriptional responses of key taxonomic groups. Front Microbiol 8:400–482

Quaiser A, Ochsenreiter T, Klenk HP, Kletzin A, Treusch AH, Meurer G, Eck J, Sensen CW, Schleper C (2002) First insight into the genome of an uncultivated crenarchaeote from soil. Environ Microbiol 4:603–611

Quaiser A, Ochsenreiter T, Lanz C, Schuster SC, Treusch AH, Eck J, Schleper C (2003) Acidobacteria form a coherent but highly diverse group within the bacterial domain: evidence from environmental genomics. Mol Microbiol 50:563–575

Riesenfeld CS, Schloss PD, Handelsman J (2004) Metagenomics: genomic analysis of microbial communities. Annu Rev Genet 38:525–552

Rochfort S (2005) Metabolomics reviewed: a new "omics" platform technology for systems biology and implications for natural products research. J Nat Prod 68:1813–1820

Rondon MR, August PR, Bettermann AD, Brady SF, Grossman TH, Liles MR, Loiacono KA, Lynch BA, MacNei IA, Minor C, Tiong CL, Gilman M, Osburne MS, Clardy J, Handelsman J, Goodman RM (2000) Cloning the soil metagenome: a strategy for accessing the genetic and functional diversity of uncultured microorganisms. Appl Environ Microbiol 66:2541–2547

Schenk PM, Carvalhais LC, Kazan K (2012) Unraveling plant–microbe interactions: can multi-species transcriptomics help? Trends Biotechnol 30:177–184

Schleper C, Jurgens G, Jonuscheit M (2005) Genomic studies of uncultivated archaea. Nat Rev Microbiol 3:479–488

Schulze WX, Gleixner G, Kaiser K, Guggenberger G, Mann M, Schulze ED (2005) A proteomic fingerprint of dissolved organic carbon and of soil particles. Oecologia 142:335–343

Sessitsch A, Weilharter A, Gerzabek MH, Kirchmann H, Kandeler E (2001) Microbial population structures in soil particle size fractions of a long-term fertilizer field experiment. Appl Environ Microbiol 67:4215–4224

Siggins A, Gunnigle E, Abram F (2012) Exploring mixed microbial community functioning: recent advances in metaproteomics. FEMS Microbiol Ecol 80:265–280

Singh BK, Campbell CD, Sorenson SJ, Zhou J (2009) Soil genomics. Nat Rev Microbiol 7:756. https://doi.org/10.1038/nrmicro2119-c1

Stewart FJ, Sharma AK, Bryant JA, Eppley JM, De Long EF (2011) Community transcriptomics reveals universal patterns of protein sequence conservation in natural microbial communities. Genome Biol 12(3):R26. https://doi.org/10.1186/gb-2011-12-3-r26

Su JQ, Xia Y, Yao HY, Li YY, An XL, Singh BK, Zhang T, Zhu YG (2017) Metagenomic assembly unravel microbial response to redox fluctuation in acid sulfate soil. Soil Biol Biochem 105:244–252

Sukul P, Schäkermann S, Bandow JE, Kusnezowa A, Nowrousian M, Leicher LI (2017) Simple discovery of bacterial biocatalysts from environmental samples through functional metaproteomics. Microbiome 64:5–28

Sun X, Zhang J, Zhang H, Ni Y, Zhang Q, Chen J, Guan Y (2010) The responses of Arabidopsis thaliana to cadmium exposure explored via metabolite profiling. Chemosphere 78:840–845

Taylor EB, Williams MA (2010) Microbial protein in soil: influence of extraction method and C amendment on extraction and recovery. Microb Ecol 59:390–399

Toplin JA, Norris TB, Lehr CR, McDermott TR, Castenholz RW (2008) Biogeographic and phylogenetic diversity of Thermoacidophilic Cyanidiales in Yellowstone National Park, Japan, and New Zealand. Appl Environ Microbiol 74:2822–2833

Topp E, Zhu H, Nour SM, Houot S, Lewis M, Cuppels D (2000) Characterization of an Atrazine-degrading Pseudaminobacter sp. Isolated from Canadian and French Agricultural Soils. Appl Environ Microbiol 66:2773–2782

Torvisk V, Ovreas L (2002) Microbial diversity and function in soil: from genes to ecosystems. Curr Opin Microbiol 5:240–245

Treusch AH, Kletzin A, Raddatz G, Ochsenreiter T, Quaiser A, Meurer G, Schuster SC, Schleper C (2004) Characterization of large-insert DNA libraries from soil for environmental genomic studies of Archaea. Environ Microbiol 6:970–980

Tringe SG, von Mering C, Kobayashi A, Salamov AA, Chen K, Chang HW, Podar M, Short JM, Mathur EJ, Detter JC, Bork P, Hugenholtz P, Rubin EM (2005) Comparative metagenomics of microbial communities. Science 308:554–557

Uga Y (2017) Genomic based ideotype breeding for root system architecture to enhance rice production. Plant Anim Genome 86:12–17

Urich T, Schleper C (2011) The "double-RNA" approach to simultaneously assess the structure and function of a soil microbial community. https://doi.org/10.1002/9781118010518.ch64

Urich T, Lanzen A, Qu J, Huson DH, Schleper C, Schuster SC (2008) Simultaneous assessment of soil microbial community structure and function through analysis of the meta-transcriptome. PLoS One 3(6):e2527. https://doi.org/10.1371/journal.pone.0002527

Utturkar SM, Bollmann A, Brzoska RM, Klingeman DM, Epstein SE, Palumbo AV, Brown SD (2013) Draft genome sequence for Caulobacter sp. strain OR37, a Bacterium tolerant to heavy metals. Genome Announc 1(3):e00322–e00313. https://doi.org/10.1128/genomeA.00322-13

Utturkar SM, Cude WN, Robeson Jr MS, Yang ZK, Klingeman DM, Land ML, Allman SL, Lu TYS, Brown SD, Schadt CW, Podar M, Doktycz MJ, Pelletier DA (2016) Enrichment of root endophytic bacteria from Populus deltoides and single cell genomic analysis. Appl Environ Microbiol 82:5698–5708

Viant MR (2009) Applications of metabolomics to the environmental sciences. Metabolomics 5:1–2

Vinaixa M, Schymanski EL, Neumann S, Navarro M, Salek RM, Yanes O (2016) Mass spectral databases for LC/MS- and GC/MS-based metabolomics: state of the field and future prospects. TrAC Trends Anal Chem 78:23–35

Wang DZ, Kong LF, Li YY, Xie ZX (2016) Environmental microbial community proteomics: status, challenges and perspectives. Int J Mol Sci 17:1275–1278

Warnecke F, Hess M (2009) A perspective: metatranscriptomics as a tool for the discovery of novel biocatalysts. J Biotechnol 142(1):91–95

Wellington EMH, Berry A, Krsek M (2003) Resolving functional diversity in relation to microbial community structure in soil: exploiting genomics and stable isotope probing. Curr Opin Microbiol 6:295–301

Willers C, Rensburg PJJ, Claassens S (2016) Can a metabolomics-based approach be used as alternative to analyse fatty acid methyl esters from soil microbial communities? Soil Biol Biochem 103:417–428

Williams MA, Taylor EB, Mula HP (2010) Metaproteomic characterization of a soil microbial community following carbon amendment. Soil Biol Biochem 42:1148–1156

Wilmes P, Bond PL (2006) Metaproteomics: studying functional gene expression in microbial ecosystems. Trends Microbiol 14:92–97

Winding A, Santos SS, Browne PD, Hansen LH, Johansen A, Krogh PH (2016) Metagenomics of bacteria, fungi and protists affected by biochar and earthworms in soil. Environ Microbiol 86:54–58

Wu L, Wang H, Zhang Z, Lin R, Zhang Z, Lin W (2011) Comparative metaproteomic analysis on consecutively Rehmannia glutinosa-monocultured rhizosphere soil. PLoS One 6:e20611. https://doi.org/10.1371/journal.pone.0020611

Xie S, Wu H, Chen L, Zang H, Xie Y, Gao X (2015) Transcriptome profiling of Bacillus subtilis OKB105 in response to rice seedlings. BMC Microbiol 15:21. https://doi.org/10.1186/s12866-015-0353-4

Yadav RK, Bragalini C, Tachet LF, Marmeisse R, Luis P (2016) Metatranscriptomics of soil eukaryotic communities. Microb Environ Genomics 1399:273–287

Yergeau E, Sanschagrin S, Beaumier D, Greer CW (2012) Metagenomic analysis of the bioremediation of diesel-contaminated Canadian high Arctic soils. PLoS One 7(1):e30058. https://doi.org/10.1371/journal.pone.0030058

Zhang J, Sun X, Zhang Z, Ni Y, Zhang Q, Liang X, Xio H, Chen J, Tokuhisa JG (2011) Metabolite profiling of Arabidopsis seedlings in response to exogenous sinalbin and sulfur deficiency. Phytochemistry 72:1767–1778

Zhou J, Deng Y, Luo F, He Z, Yang Y (2011) Phylogenetic molecular ecological network of soil microbial communities in response to elevated CO_2. MBio 2(4):e00122–e00111. https://doi.org/10.1128/mBio.00122-11

Zyśko A, Sanguin H, Hayes A, Wardleworth L, Zeef LAH, Sim A, Paterson E, Singh BK, Kertes MA (2012) Transcriptional response of Pseudomonas aeruginosa to a phosphate-deficient Lolium perenne rhizosphere. Plant Soil 359:25–44

Diversity of Sulfur-Oxidizing and Sulfur-Reducing Microbes in Diverse Ecosystems

4

Upendra Kumar, P. Panneerselvam, Vadakattu V. S. R. Gupta, M. Manjunath, Priyanka Priyadarshinee, Archana Sahoo, Soumya Ranjita Dash, Megha Kaviraj, and K. Annapurna

Abstract

Sulfur (S) is one of the most important elements, of which the organosulfur compounds and/or metal sulfides are considered essential for life. Microbial sulfur oxidation and reduction are the most active and ancient metabolic processes in S cycle that operate in diverse ecosystems. This process is carried out by sulfur-oxidizing (SOB) and sulfur-reducing bacteria (SRB) in all ecosystems and considered as key phenomenon in sulfur biogeochemical cycling. Usually, on the basis of nutrition, SOB and SRB are categorized as lithoautotrophs. SOB oxidize the reduced sulfur compounds such as hydrogen sulfide (H_2S), elemental sulfur (S^0), sulfite (SO_3^{-2}), thiosulfate ($S_2O_3^{2-}$), and various polythionates ($S_nO_6^{2-}$ or -S_nO_6-) into sulfate (SO_4^{-2}). On the contrary, SO_4^{-2} can serve as an electron acceptor of SRB under anaerobic condition, and they reduce the SO_4^{-2} and other oxidized sulfur compounds ($S_2O_3^{2-}$, SO_3^{-2}, S^0) into H_2S. In natural system, SRB reduce the SO_4^{-2} in two different reduction processes, *viz*, dissimilatory and assimilatory reactions. In dissimilatory reaction, SRB utilize three kinds of enzymes (ATP sulfurylase, APS reductase, and sulfite reductase) to reduce the S substrate, whereas the sulfate is assimilated or incorporated into organic compounds under assimilatory process through S substrate reduction. In recent years, molecular methods have emerged as essential tools for a better

U. Kumar (✉) · P. Panneerselvam · P. Priyadarshinee · A. Sahoo · S. R. Dash · M. Kaviraj
ICAR-National Rice Research Institute, Cuttack, Odisha, India
e-mail: ukumarmb@gmail.com; ukumarmb@yahoo.co.in; upendra.kumar1@icar.gov.in

V. V. S. R. Gupta
Commonwealth Scientific Industrial Research Organization (CSIRO), Adelaide, Australia

M. Manjunath
ICAR-Central Research Institute for Dryland Agriculture (CRIDA), Hyderabad, Telengana, India

K. Annapurna
Division of Microbiology, ICAR-Indian Agricultural Research Institute, New Delhi, India

T. K. Adhya et al. (eds.), *Advances in Soil Microbiology: Recent Trends and Future Prospects*, Microorganisms for Sustainability 3,
https://doi.org/10.1007/978-981-10-6178-3_4

understanding of the microbial role in S transformation under various habitats. Keeping the importance of microbial-mediated S oxidation and reduction in biogeochemical cycle of S, the present chapter describes the role of key functional microbial genes in S transformation such as genes involved in S oxidation (*sox*, *aps*, *asf*, and *sor*) and reduction (*dsr*) and also discusses in detail about the abundance, diversity, and impact of these in diverse ecosystems.

Keywords
Sulfur · Oxidation · Reduction · Functional genes · Microbial diversity

4.1 Introduction

Sulfur (S) is the tenth most copious element in the universe and the sixth most prominent element in microbial biomass (Klotz et al. 2011). It is present throughout the earth's crust as gypsum and pyrite. Sulfur comes from weathered rock, atmosphere (SO_2 and methane sulfonic acid), fertilizers and pesticides, water resources (sulfate, hydrogen sulfide, and elemental sulfur), etc., and these processes are influenced by climate, local vegetation, and topography. The sulfur content of soil varies from 0.002 to 10.0% (Freney et al. 1982), and the highest amount of S is present in tidal flats, saline, acid sulfate, and organic soils. Organic S accounts to >90% of total sulfur present in surface soils, whereas <25% of total S present in agricultural soils are in the form of the inorganic S (Roberts and Bettany 1985; Bettany et al. 1973). The main forms of inorganic sulfur include sulfide, elemental sulfur, sulfite, thiosulfite, tetrathionate, and sulfate (Williams 1972).

Majority of global biogeochemical cycles including that of carbon, nitrogen, phosphorus, iron, and sulfur are driven by microorganisms (Tang et al. 2007). Approximately one-half of the global S cycle represents oxidized form of inorganic S compounds. The bacteria and archaea responsible to form oxidized form of S from reduced S compounds, belonged to either photolithotrophs or chemolithotrophs (Trüper and Fischer 1982; Brune 1989; Takakuwa et al. 1992; Nelson and Fisher 1995; de Zwart et al. 1996; Kelly et al. 1997; Friedrich et al. 2001). Under photolithotrophic growth, green and purple sulfur-oxidizing bacteria (SOB) utilize S compounds as electron donors for reductive carbon dioxide fixation (Brune 1989; Brune et al. 1995), and light energy is used as electrons transfer from S compounds *via* highly reducing electron carriers such as NAD (P) and ferredoxin. This process includes a wide range of enzymes, involved in catalyzing sulfur redox reactions (Trüper and Fischer 1982; Fischer 1989; Brune 1989; Dahl and Truper 1994; Brune et al. 1995). In sulfide oxidation, oxidation of elemental S is catalyzed by sulfide dehydrogenase initially and then catalyzed by flavocytochrome c, other c-type cytochromes, or sulfide/quinone oxidoreductase (Brune 1989; Brune et al. 1995). Another enzyme, siroheme sulfite reductase, oxidized H_2S directly to sulfite, and this enzyme is found in *Chromatium vinosum* D (Schedel et al. 1979). Besides photolithotrophs, reduced sulfur compounds such as hydrogen sulfide (H_2S), elemental sulfur (S^0), sulfite (SO_3^{-2}), thiosulfate ($S_2O_3^{2-}$), and various polythionates ($S_nO_6^{2-}$ or -S_nO_6-) are utilized by various chemolithotrophs, and they oxidized these forms of S into sulfate (SO_4^{-2}). SO_4^{-2} can serve as an electron acceptor in anaerobic respiration, and S-reducing bacteria (SRB) may reduce the SO_4^{-2} and other oxidized sulfur compounds (S^0, SO_3^{-2}, and $S_2O_3^{2-}$) into H_2S.

Sulfur is essential for the growth and development of living organisms. Plants require it for growth and grain production. Plants generally utilize S in the form of SO_4^{-2}. Due to its existence as several redox states, S is involved in very important biochemical reactions as redox center and carbon carrier (Klotz et al. 2011). Therefore, it is very important to know the nature and amount of S that is present in soil and its transformation process. The main purpose of this chapter is to briefly introduce to researchers how the S cycle is mediated through microbes under different ecosystems and also discuss the kind of functional genes required in S transformation and their abundance, diversity, and impact in diverse ecosystems.

4.2 Biogeochemical of Sulfur Cycle

Sulfur (S) is considered as one of the most important atoms in biological system, and minute amount of this element is mandatory for proper functioning of biological system. Generally, S forms disulfide bridges in biological system, which imparts crucial function to provide specific shapes and properties to other biologically important molecules under that system. The major constituent of S present in the atmosphere is sulfur dioxide (SO_2), coming from burning of fossil fuels and sulfur coal. In the atmosphere, one of the primary components of acid rain is sulfuric acid (H_2SO_4) which is formed when SO_2 is reacted with water vapor and causing many adverse effects in about all regions of the world. Carbonyl sulfide (COS) is another form of S, present in small quantity in the atmosphere. These two forms of S (SO_2 and COS) are highly reactive with oxygen and converted into sulfates (SO_4^{-2}) which is quickly deposited on land and other surfaces. Plant requires S in the form of SO_4^{-2}, and these forms of S mostly come from soil organic S after mineralization by soil microorganisms (Niknahad-Gharmakher et al. 2012). Greater availability of soil S may be immobilized by soil microbes to build their biomass. In soil, the S level in the form of extractable soil S-SO_4 is marginally changed after C addition over time, confirming slow rate of soil S mineralization. Majority (90–95%) of soil S is stored in organic form such as C-S (sulfate bonded with carbon) and/or C-O/N-S (sulfate and sulfamates in the form of ester) (Tabatabai 1984). Plant generally takes sulfates from the ester sulfate fractions (McLaren et al. 1985). In agrosystems, plant residues are the main source of labile carbon (Gentile et al. 2011). The effect of plant residues on S turnover in soils has been studied and found that the net S mineralization was the function of C/S ratio of the crop residue (Jensen et al. 2005). This gives an idea that the S deficiency in cropped soil is functionally correlated with soil microbial biomass and C dynamics (Wu et al. 1995). One of the key factors governing S transformations in soil is availability of C (Knights et al. 2001); probable limiting effect of low S levels on C mineralization has been investigated by few workers

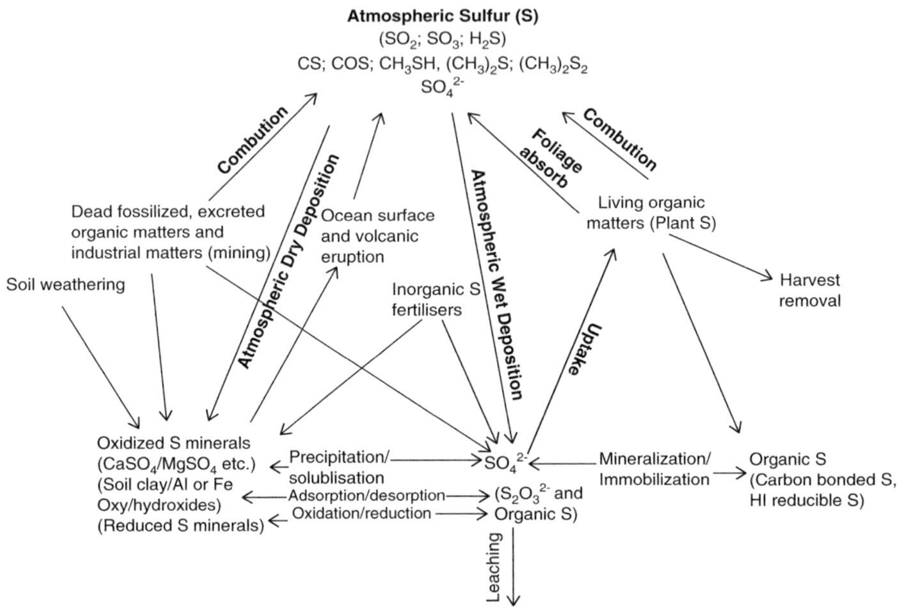

Fig. 4.1 Representative model of biogeochemical cycle of sulfur (Adapted from Germida et al. 1992)

(Chapman 1990). Moreover, Fig. 4.1 presents a representative S cycle which showed the transformation of different forms of S, its compounds, and their metabolic reaction under model system.

4.3 Microbiology of Sulfur Cycling

Biogeochemical cycle of S involves several oxidation and reduction reactions (Tang et al. 2007). The following major pathways involved in S cycle are (1) organic S mineralized into inorganic S form; (2) oxidation of S^0, SO_3^{-2}, and $S_2O_3^{2-}$ into SO_4^{-2}; (3) reduction of SO_4^{-2} into H_2S; and (4) immobilization of S compounds by microbes and subsequent assimilation of S as organic form (http://www4.ncsu.edu). Under S cycling, microorganisms can take part in both oxidation and reduction processes depending on the prevailing environmental conditions in a particular ecosystem.

4.3.1 Sulfur Oxidation by Microbes

Sulfur (S) oxidation is one of the most predominant reactions in environment (Friedrich et al. 2005). It is a very significant process in soil to avoid sulfur deficiencies in crops and also the environmental contamination (Lawrence et al. 1988). The archaea (*Sulfolobus*, *Desulfurococcus*, *Acidianus*, *Metallospora*), chemolithotrophic bacteria (*Bacillus*, *Acidithiobacillus*), phototrophic bacteria (*Chlorobium*, *Allochromatium*, *Rhodobacter*, *Rhodovulum*), and non-sulfur bacteria (*Rhodopseudomonas*, *Rhodocyclus*) are mainly involved in S oxidation. The majority of them use carbon dioxide as their primary carbon source and sulfur as an electron donor (Brune 1989; Friedrich et al. 2005). The sulfur substrates utilized by the microorganisms include sulfide, thiosulfate, and hydrogen sulfide (Friedrich et al. 2001, 2005). The process of S oxidation occurs through three biochemical pathways: sulfur oxidase pathway, the reverse siroheme sulfite reductase pathway (phototrophic S-oxidizing bacteria), and the archaeal sulfur oxygenase reductase pathway.

4.3.1.1 Phototrophic Sulfur Bacteria

Phototrophic S-oxidizing bacteria use light as energy source and hydrogen sulfide (H_2S) as substrate. They usually oxidized H_2S into elemental sulfur (S^0) and subsequently reduce the carbon dioxide and assimilated as organic compounds. There are several enzymes involved for catalyzing sulfur redox reactions in phototrophic sulfur bacteria (Trüper and Fischer 1982; Fischer 1989; Brune 1989; Dahl and Truper 1994; Brune et al. 1995), which are discussed below with examples.

Green Sulfur Bacteria

Green sulfur bacteria (GSB) are metabolically strict anaerobes and obligately phototrophic and use CO_2 as only carbon source and fixed *via* the reductive tricarbonic acid cycle. Sulfide (H_2S) is used as electron (e^{-1}) donor by all species of GSB except *Chlorobium ferrooxidans* (iron-oxidizing GSB) and subsequently oxidized to sulfate (SO_4^{-2}) with intermediary assimilation of extracellular S. Many are able to grow with elemental S, and even some species also use thiosulfate ($S_2O_3^{2-}$) (Frigaard and Bryant 2008). Tetrathionate may be used as electron donor in some of the GSB like *Chlorobaculum parvum* and *Chlorobium thiosulfatophilum* (Imhoff 2003; Khanna and Nicholas 1982; Larsen 1952). So far, sulfite (SO_3^{-2}) utilization has not yet been discovered in the case of any GSB. Some of the most important GSB are *Chlorobium*, *Ancalochloris*, *Pelodictyon*, *Chloroherpeton*, etc.

Purple Sulfur Bacteria

Purple S-oxidizing bacteria (PSOB) generally use sulfide (H_2S) for their growth and development. They store the sulfur in the form of spherical particles within and outside of the cells and upon oxidation releases sulfates from the cells. They oxidize the sulfide, sulfur, thiosulfate, and sulfite (Imhoff and Hiraishi 2005) to sulfate by different mechanisms. PSOB have two different kinds of pathways for thiosulfate oxidation. In one pathway, two thiosulfate anions were oxidized by enzyme thiosulfate dehydrogenase and produce tetrathionate, whereas another pathway sulfate, was produced after complete oxidation of thiosulfate (Dahl and Friedrich 2008). *Chromatium*, *Allochromatium*, *Thiocystis*, *Thiococcus*, *Thiospirillum*, etc. are the known PSOB in natural environment.

Purple Non-sulfur Bacteria

The occurrence of purple non-sulfur bacteria (PNSB) is wider and heterogeneous, belonging to photoautotrophs which use hydrogen (H_2) or sulfide (H_2S) as electron donor. Some groups of PNSB do not oxidize H_2S completely to sulfate (SO_4^{-2}); instead they form sulfur (S) as an end product. However, SO_4^{-2} is the final end product in the H_2S mediated by many PNSB such as *Rhodovulum*, *Rhodopseudomonas palustris*, *Blastochloris sulfoviridis*, etc. (Brune et al. 1995; Imhoff and Hiraishi 2005). While thiosulfate is oxidized into tetrathionate by *Rhodopila globiformis* (Then and Trüper 1981), *Rhodovulum* species oxidize thiosulfate completely into SO_4^{-2} (Brune et al. 1995; Appia-Ayme et al. 2001; Imhoff and Hiraishi 2005). Most of the PNSB may grow as chemoorganotrophs under microoxic to oxic conditions without presence of light (Smith and Lascelles 1966; Trüper and Pfennig 1966).

4.3.1.2 Chemolithotrophic Sulfur-Oxidizing Bacteria

Chemolithotrophic sulfur-oxidizing bacteria (CSOB) use reduced inorganic sulfur compounds such as sulfite, thiosulfate, hydrogen sulfide, etc. as their energy source. There are two major groups: (1) the obligate chemolithotrophic bacteria, which usually receive energy from the oxidation of S and use main carbon source as CO_2, and (2) the facultative autotrophic bacteria, or mixotrophic bacteria, which can grow autotrophically, mixotrophically, or even heterotrophically. The chemolithotrophic bacteria such as *Thiobacillus ferrooxidans* and *T. thiooxidans* are commonly present bacteria generally responsible for S^0 oxidation in soils and also considered as the most important precursor for S-biogeochemical cycle. *Thiobacillus thiooxidans* is a chemolithotrophic acidophilic bacterium that uses S^0 as an energy source and is important in the microbial catalysis of H_2S. However, the significant number of *Thiobacillus* is not reported in most of the agricultural soils (Chapman 1990; Lawrence et al. 1988; Tourna et al. 2014; Zhao et al. 2017a). *Beggiatoa leptomitiformis* is also a CSOB which uses succinate and thiosulfate or tetrathionate and grows as mixotrophs and oxidized substrate to generate ATP by oxidative phosphorylation. Some of the common CSOB are *Thiobacillus*, *Thiothrix*, *Beggiatoa*, etc.

4.3.1.3 Autotrophic Denitrifying Sulfur-Oxidizing Bacteria

Autotrophic denitrifying sulfur-oxidizing bacteria (ADSOB) generally use various reduced sulfur compounds and produce nitrogen gas by the reduction of nitrate or nitrites. Some of the common ADSOB are *Thiobacillus denitrificans*, *T. versutus*, *Thiosphaera pantotropha*, *Pseudomonas denitrificans*, etc.

4.3.1.4 Heterotrophic Sulfur-Oxidizing Microbes

Heterotrophic sulfur-oxidizing bacteria (HSOB) could oxidize sodium sulfide, tetrathionate, thiosulfate, metabisulfite, and sulfite, but they are unable to gain energy from S oxidation (Tuttle 1980). Starkey (1934) confirmed that HSOB isolated from soil could oxidize $S_2O_3^{2-}$ both in organic and mineral media, with $S_4O_6^{2-}$ being formed as an intermediate. HSOB could also oxidize $S_4O_6^{2-}$ to tri-

and pentathionate, and these oxidations being associated with an initial rise and then a fall in the pH of the culture medium suggest that the growth of some heterotrophic marine bacteria is stimulated when $S_2O_3^{2-}$ is oxidized. A range of hydrogen bacteria (*Xanthobacter autotrophicus*, *Aquaspirillum autotrophicum*, *Pseudomonas pseudoflava*, and *P. pulleronii*) was shown by Friedrich and Mitrenga (1981) to be capable of oxidizing $S_2O_3^{2-}$. Hydrogen sulfide (H_2S)-oxidizing actinomycetes isolated from soil could oxidize S as facultative chemoautotrophs. However, these organisms also act as heterotrophs and are able to scavenge carbon from the atmosphere (Skiba and Wainwright 1984). To date the list of fungi capable of S oxidation contains mainly soil fungi such as *Asteriomyces crucicatus*. Thermophilic fungus *Sporotrichum thermophile* can oxidize SO to $S_2O_3^{2-}$ at 37 to 45 °C. Even ectotrophic mycorrhizae can play a vital role in sulfur oxidation in soils. *Aspergillus niger* and *Mucor fiaous* oxidized elemental sulfur in vitro to form relatively large amounts of sulfate. Some of the examples of HSOB are *Pseudomonas aeruginosa*, *Sphaerotilus natans*, *Xanthobacter autotrophicus*, *Aquaspirillum autotrophicum*, *Pseudomonas pseudoflava*, *P. pulleronii*, *Actinomycetes*, *Alternaria tenuis*, and *Aureobasidium pullulans*. A soil amoeba has been shown to be capable of oxidizing H_2S.

4.4 Sulfur Reduction by Microbes

Microbial sulfur (or sulfate) reduction is governed by two possible pathways, i.e., either assimilatory or dissimilatory process. In the assimilatory reduction pathway, reduced sulfur is generally used for biosynthesis of amino acids and proteins, whereas in dissimilatory reduction, sulfate (or sulfur) is reduced to inorganic sulfide by obligatory anaerobic sulfate reducers. The process of sulfur reduction occurs through dissimilatory sulfur reductase system which is present both in bacterial and archaeal sulfate-reducing species (Wagner et al. 1998). The organisms which are involved in this process draw majority of their metabolic energy from the reduction and use of sulfur compounds as electron acceptors. In this process, carbon substrates such as lactate or ethanol are oxidized, and hydrogen sulfide gas is produced (Jørgensen 1982). The enzyme pathway responsible for the reduction of sulfur is known as the dissimilatory sulfur reductase system. The sulfur-reducing organisms (SRB) are generally found in anaerobic conditions and play vital role in the formation of acid sulfate soils and pyrite. Sulfide can be produced by anaerobic microorganisms while breaking proteins to amino acids. Some of the examples of SRB and archaea are *Desulfurella*, *Desulfuromonas*, *Geobacter*, *Pelobacter*, etc. and *Thermoproteales*, *Thermococcales*, *Sulfolobales*, *Pyrodictales*, *Sulfolobales*, etc., respectively (Schauder and Kröger 1993).

4.5 Microbial Functional Genes Responsible for Sulfur Oxidation

In the recent years, sulfur oxidation pathways have been reported in many S-oxidizing bacteria (SOB), and the biochemistry behind these pathways is quite complicated (Ghosh and Dam 2009). In general, SOB follow two types of S oxidation pathways; one is Sox pathway (*sox* gene) which involves a multienzyme complex catalyzing the complete oxidation of reduced sulfur compounds to sulfate, and another is APS (adenosine-5-phosphosulfate) pathway (*aps* gene) which implements elemental sulfur and sulfite as intermediates (Ghosh and Dam 2009). Other important genes in S oxidation pathway are *asf* and *sor*. *Asf* gene is responsible for aryldesulfonation reaction of sulfonate mostly present in agricultural soils, whereas *sor* gene encodes sulfur oxygenase reductase, which oxidized the elemental sulfur and produced sulfite, thiosulfate, and sulfide. Comprehensive information of function of various key genes associated with biogeochemical cycle of sulfur is presented in Fig. 4.2.

4.5.1 *sox* Gene

The Sox (sulfur oxidase pathway) is currently considered the most widely distributed and the best characterized of the bacterial and archaeal S oxidation pathways. The S*ox* enzyme pathway is responsible for the oxidation of reduced S or S compounds and has been isolated in polythionate-oxidizing bacteria (Bamford et al. 2002). Common sulfur oxidase enzymology in the bacteria was initially illustrated by Trüper and Fischer (1982) in a comparison of chemoautotrophic and phototrophic bacteria. It was noted that a number of enzymes were common to the green, purple, and colorless sulfur bacteria, including the common use of cytochrome C and flavocytochrome C in electron transport (Trüper and Fischer 1982; Friedrich et al. 2001). The Sox enzyme system was originally classified as a number of separate pathways. Each of the pathways was designated principally by function, most

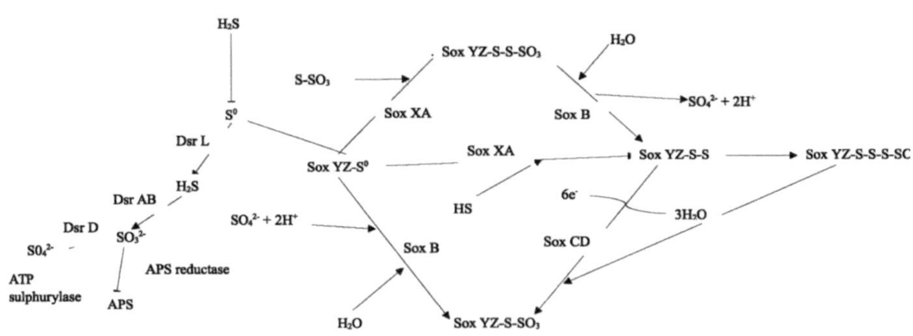

Fig. 4.2 Function of various key genes associated with biogeochemical cycle of sulfur (Adapted from Grabarczyk et al. 2015)

commonly thiosulfate oxidation, due to both the stability of the thiosulfate molecule and the common utilization of thiosulfate by the majority of the bacteria (Petri et al. 2001).

Sox complex has many components such as *soxB*, *soxXA*, *soxYZ*, and *soxCD*. The key constituent among all is *soxB*. The oxidation of thiosulfate ($S_2O_3^{2-}$) to form sulfate (SO_4^{2}) is stringently dependent on the presence of three periplasmic Sox proteins which has been encoded by *soxBXA* and *soxYZ* genes. However, Sox proteins are not necessarily required during oxidation of sulfide (H_2S) process (Hensen et al. 2006). Purple sulfur bacteria comprise 15 different kinds of *sox* genes which have been organized into three transcriptional units such as *soxRS*, *soxVW*, and *sox XYZABCDEFGH*. Out of these, *in vivo* and *in vitro* thiosulfate oxidation are essentially mediated by periplasmic proteins SoxXA, SoxYZ, SoxB, and Sox $(CD)_2$. In green S bacteria (*Chlorobaculum parvum* DSM 263), *soxJ-soxXYZA-soxK-soxBW* genomic arrangement is generally found (Frigaard and Bryant 2008) which forms sulfur (S) during thiosulfate oxidation (Steinmetz and Fischer 1982). Polysulfides may act as intermediates during thiosulfate oxidation in the periplasm of green sulfur bacteria (Frigaard and Bryant 2008; Friedrich et al. 2001). Green S bacteria, *Allochromatium vinosum*, lack the enzyme sulfur dehydrogenase; therefore the sulfane sulfur atom which is linked to *soxY* cannot be oxidized. However, other genes *soxB* and *soxXA* are transcribed divergently in *A. vinosum* (Frigaard and Bryant 2008). Among all *sox* genes, *soxCD* gene is not detected in magnetotactic *Magnetococcus* sp. MC1, *Thiobacillus denitrificans*, thiosulfate-oxidizing green sulfur bacteria, and *A. vinosum* (Frigaard and Bryant 2008).

4.5.2 *aps* Gene

Adenosine-5-phosphosulfate (APS) pathway involves two enzymes such as APS reductase and ATP sulfurylase (Kappler and Dahl 2001). APS reductase is encoded by *aps* gene which forms APS after catalyzing sulfite and adenosyl monophosphate (AMP) during indirect sulfite oxidation. ATP sulfurylase (ATP, sulfate adenyltransferase) and adenylsulfate/phosphate adenyltransferase (APAT) catalyze to transfer AMP moiety of APS to either pyrophosphate or phosphate, respectively. APS reductase also acts as key enzyme in dissimilatory sulfate reduction pathway in sulfur-reducing prokaryotes (Meyer and Kuever 2007). However, this enzyme is involved in the transformation of sulfite to APS in sulfur-oxidizing prokaryotes (Meyer and Kuever 2007). The *aps* gene was first identified in the archaea *Acidianus ambivalens* in which the major enzyme, sulfur oxygenase reductase, catalyzes the oxidation of sulfur (Urich et al. 2005). Recently, this enzyme system has also been detected in multiple members of the bacteria including *Acidithiobacillus* species and *Aquifex aeolicus*.

4.5.3 *asfA* Gene

Assimilation and mobilization sulfonates in agricultural soils are one of the key soil processes in S cycle, and this is mediated by microbial oxidoreductase *asfA* gene. The *asfA* gene was first discovered in *Pseudomonas putida* S-313, which has the ability to desulfurize toluene sulfonate to p-cresol under aryldesulfonation process (Vermeij et al. 1999; Kertesz and Mirleau 2004). Orthologue sequences of *asfA* gene are detected in vast group of cyanobacteria and bacteria including *Cupriavidus* (*Ralstonia*) *metallidurans* which are able to utilize arylsulfonates as sulfur source. A 100-fold increase in the expression of *asfA* gene was detected in *C. metallidurans* or *P. putida* S-313 culture media containing toluene sulfonate as sulfur source, but the expression was largely repressed when sulfate was added. Kertesz and Mirleau (2004) analyzed the *asfA* containing bacterial diversity in barley rhizosphere and indicated the huge diversity of bacteria that were capable to utilize toluene-sulfonate as sulfur source.

4.5.4 *sor* Gene

The sulfur oxygenase reductase (Sor) enzyme is encoded by *sor* gene which oxidizes the elemental sulfur into sulfite and thiosulfate. The Sor enzyme is generally considered as "archaeal-like" enzyme and present in acidophilic leaching bacteria such as *Acidithiobacillus caldus*, *A. thiooxidans*, *A. ferrivorans*, and *Sulfobacillus thermosulfidooxidans* (Janosch et al. 2015). Sor is a thermophilic enzyme, and its oxygenase activity was detected at 75 °C in *Sb. thermosulfidooxidans* DSM 9293T. Besides *sor* genes, oxygenase activity in *Sb. thermosulfidooxidans* DSM 9293T also has another kind of genes which encodes complete heterodisulfide reductase (*hdr* gene), tetrathionate hydrolase (*tth* genes), sulfide/quinone reductase (*sqr gene*), and thiosulfate quinone reductase (*tqo*) gene. Interestingly, no *sox* genes were involved in the oxygenase activity.

4.6 Microbial Genes Involved in Sulfur Reduction

4.6.1 *dsr* Gene

Sulfate-reducing bacteria (SRB) contain *dsr* gene which encodes the dissimilatory sulfite reductase and is able to catalyze the conversion of sulfite to sulfide with reduction of six electrons. Different models have been proposed to explain the exact roles of the *dsr*-encoded proteins in *Allochromatium vinosum* (Dahl et al. 2005). Altogether, 15 open reading frames, designated *dsrABEFHCMKLJOPNRS*, were identified in *A. vinosum* (Hipp et al. 1997; Lübbe et al. 2006). Various studies have been carried out to study the diversity of SRB using a 1.9-kb *dsrAB* gene fragment amplified with DSR1F and DSR4R primers. These primers were used for molecular characterization of SRB from various habitats including deep sea hydrothermal vents, salt marshes, sediments, etc. (Agrawal and Lal 2009).

4.7 Microbial Association in Sulfur Cycle Under Diverse Ecosystems

The representative microbial groups responsible for sulfur oxidation and reduction in different ecosystems are elucidated in Table 4.1.

4.7.1 Agroecosystems

The impact of sulfur (S) deficiency in agriculture soils has been recognized for more than a century and is becoming increasingly common in many areas of the world as a result of intensive agriculture, high biomass exportation, and reduced S emissions to the atmosphere (Lucheta and Lambais 2012). Among agricultural crops, rice is the dominant and staple food crop of Asia having 90% of the world's total rice grain production. As rice plants can occupy a large volume of the planted soil, oxidized zones can occur which allow the growth and metabolism of aerobic microorganisms, even in flooded conditions (Freney et al. 1982). As a result, sulfur can exist in these soils in all of its oxidation states from +6 of sulfate to −2 of sulfide, and reduced forms of the element are subject to normal oxidation processes, although sulfur oxidation in paddy soils has not been studied extensively. The two microbes, *Thiobacillus thioparus* (Freney et al. 1982) and *T. thiooxidans* (Mouraret and Baldensperger 1977) have been isolated, and other species are likely to be present (Freney et al. 1982). It has already been mentioned that *Beggiatoa* species (Joshi and Hollis 1976) play a dominant role in rice soils, and it is also likely that heterotrophs and purple and green sulfur bacteria are important in the oxidation of reduced S in the rice rhizosphere. It has been reported that the oxidation of sulfide is beneficial for the rice growth and H_2S served as a causal agent in 12 out of the 27 physiological disorders of rice. On the other hand, soluble sulfides are toxic to nematodes and, hence, can be beneficial to rice (Freney et al. 1982).

As sulfur deficiencies are coming up in rice growing, making necessary sulfur fertilization with compounds such as elemental sulfur and sulfur-coated urea, there is a clear need for a better understanding of the sulfur oxidation in rice paddy soils. Reductions of sulfate, under paddy soil, play key roles in the nutrient mineralization process under early flooded rice fields (Yao et al. 1999). Researchers indicated that sulfur concentration is slightly lower in rice field flooded with freshwater than the marine ecosystem. Another study suggested that soil incorporated with rice straw significantly increased sulfate content. The sulfate (SO_4^{2-}) reduction was observed higher in the rice straw-amended slurries due to presence of high *dsrAB* gene copy numbers. Most of the bacteria responsible for SO_4^{2-} reduction in this condition belonged to the genera *Clostridia*, *Desulfobacterium*, *Desulfovibrio*, *Desulfomonile*, and *Syntrophobacter* (He et al. 2010). Recent study by Kumar et al. (2017) revealed

Table 4.1 Association of different sulfur-oxidizing and sulfur-reducing microbes in various ecosystems

Habitat	Microbes	Response (oxidation/ reduction)	References
Agriculture ecosystem	*Beggiatoa sp.* (paddy soil)	Oxidation	Burke et al. (1974) and Joshi and Hollis (1976)
	Thiobacillus denitrificans (cotton and groundnut field)	Oxidation	Yousuf et al. (2014)
	T. thioparus	Oxidation	Wainwright (1984)
	T. neapolitanus	Oxidation	Wainwright (1984)
	T. novellus	Oxidation	Wainwright (1984)
	Rhodovulum sulfidophilum	Oxidation	Yousuf et al. (2014)
	Betaproteobacteria	Oxidation	Yousuf et al. (2014)
	Marichromatium purpuratum	Oxidation	Yousuf et al. (2014)
Aquatic ecosystem	*Beggiatoa sp.*	Oxidation	Wainwright (1984)
Barren terrestrial land ecosystem			
	Rhodothalassium salexigens	Oxidation	Yousuf et al. (2014)
	Thiomicrospira crunogena	Oxidation	Yousuf et al. (2014)
	Paracoccus pantotrophus	Oxidation	Bardischewsky et al. (2005)
Blacks Drain and Cudgen Lake	*Aquifex aeolicus*	Oxidation	Pelletier et al. (2008)
	Paracoccus versutus	Oxidation	Wodara et al. (1997)
	Archaeoglobus profundus	Reduction	Mander et al. (2004)
	Thermodesulforhabdus norvegica	Reduction	Larsen et al. (2001)
	Desulfotomaculum thermocisternum	Reduction	Larsen et al. (2001)
Coastal saline land, hypersaline habitats	*Rhodovulum sulfidophilum*	Oxidation	Tourova et al. (2011)
	Thiomicrospira crunogena	Oxidation	Tourova et al. (2011)
	Spirochaeta sp.	Oxidation	Tourova et al. (2011)
	Rhodovillum adriaticum	Oxidation	Tourova et al. (2011)
Coastal acid sulfate soil under sugarcane cultivation	*Acidithiobacillus ferrooxidans*	Oxidation	Wakai et al. (2004)
Costal ecosystem	*Thiomicrospira sp.*, *Arcobacter sulfidicus*, and *Sulfurimonas denitrificans*	Oxidation	Kuenen and Tuovinen (1981)
Freshwater ecosystem	*Betaproteobacteria*	Oxidation	Wu et al. (2006)

(continued)

Table 4.1 (continued)

Habitat	Microbes	Response (oxidation/ reduction)	References
Hot spring ecosystem	*Proteobacteria*	Reduction	Badhai et al. (2014)
	Thermodesulfovibrio sp.	Reduction	Badhai et al. (2014)
	Thiobacillus ferrooxidans	Oxidation	Wainwright (1984)
	Thiobacillus organoparus	Oxidation	Wainwright (1984)
	Mycorrhizae	Oxidation	Grayston and Wainwright (1988)
Hypersaline habitats	*Thiohalorhabdus denitrificans*	Oxidation	Sorokin et al. (2008)
Lihir Island	*Acidianus sulfidivorans sp. nov.*	Oxidation	Plumb et al. (2007)
Mangrove swamps	*Desulfovibrio desulfuricans*	Reduction	Sahoo and Dhal (2009)
Marine ecosystem	*Asteriomyces crucicatus*	Oxidation	Wainwright (1984)
	Asteriomyces crucicatus	Oxidation	Wainwright (1984)
	Oscillochloris trichoides	Oxidation	Dahl and Friedrich (2008)
Marine sediments	*Gammaproteobacteria*	Oxidation	Yousuf et al. (2014)
	Archaeoglobus fulgidus	Reduction	Mander et al. (2004)
	Thioploca sp.	Oxidation	Jørgensen and Nelson (2004)
	Pseudoxanthomonas mexicana	Oxidation	Krishnani et al. (2010)
Spruce forest ecosystem	*T. thiooxidans*	Oxidation	Wainwright (1984)
	T. thioparus	Oxidation	Wainwright (1984)
Sub-tropical rainforest and back swamps	*Ferrobacillus ferrooxidans*	Oxidation	Brunner et al. (2008)
Swamp ecosystem	*Aspergillus niger*	Oxidation	Grayston et al. (1986)
	Mucor fiaous	Oxidation	Grayston et al. (1986)

that the temporal variation of sulfur-oxidizing bacteria (SOB) was observed under continuous application of chlorpyrifos over seven seasons in paddy soil.

Canola plant (*Brassica napus*) requires high sulfur (S) during its vegetative growth; otherwise, it shows S-deficiency symptoms. Therefore, elemental sulfur (S^o) fertilizer (with or without inoculated sulfur-oxidizing microorganisms) is frequently used to alleviate this problem (Anandham 1991). *Burkholderia* sp. strain ATSB13T, a thiosulfate-oxidizing facultative chemolithoautotrophic, was isolated from tobacco rhizosphere and has ability to serve as a potential inoculant along with elemental sulfur fertilizers (Anandham et al. 2009).

4.7.2 Acid Sulfate Soil

Acid sulfate soils (ASS) are widespread around the globe and are formed by natural accumulation of bacterially formed pyrite in estuarine environments such as mangrove swamps (White and Engelen 1997) worldwide. ASS is the name given to all soils and sedimentary materials that, through pedogenesis, produce sulfuric acid in quantities that affect soil properties. Southeast Asia occupied about half of the area of ASS found in the world (Langenhoff 1986). In India, these soils are mostly located in swampy coastal plains in the Kuttanad tract (kari lands) of Kerala (Mathew et al. 2001). Alteration of soil water regimes has occurred following the increased urban and rural development of coastal regions. The subsequent oxidation of metal sulfide materials in these soils generates sulfuric acid and highly acidic soil conditions (Dent 1986). ASS sites release leachate of low pH metal which is one of the factors responsible for severe contamination and degradation of ecosystem. As such, the oxidation of ASS results in a host of environmental and economic problems that include loss of aquatic habitats and populations, decreased soil productivity, the emission of greenhouse and other gasses into the atmosphere, and the degradation of civil infrastructure. ASS oxidation also reduces the productivity of agricultural land and decreases the ecological health of aquatic ecosystems through the release of acidic leachate. It is proposed that bacterial and archaeal communities play an important role in the oxidation of ASS and the subsequent generation of acid similar to those observed in acid mine drainage environments.

ASS oxidation means oxidation of pyrite which produces a wide range of oxidation products including sulfuric acid. There is a number of oxidation pathways described for the complete oxidation of pyrite. The complete oxidation of pyrite is proposed to proceed via the formation of intermediates including elemental sulfur. A two-step oxidation then produces ferrous iron (Fe^{2+}) and sulfate followed by further oxidation to produce ferric iron (Fe^{3+}). This oxidation process has been referred to as ripening of ASS (Dent 1986). Ferric iron (Fe^{3+}) has the capacity to oxidize pyrite directly in an oxygen-independent reaction. This interaction can further accelerate the oxidation process. Sometimes, ASS oxidation occurs naturally, as a result of drought and increased pressure on groundwater supplies (lower water table elevation) due to evapotranspiration. The oxidation process is often balanced by natural re-flooding events, which reduce the severity and impacts of oxidation products. Biologically mediated pyrite oxidation is attributed exclusively to the activity of bacteria, and acidophilic chemolithotrophic bacteria such as *Acidithiobacillus ferrooxidans* and *Ferrobacillus ferrooxidans* are responsible to catalyze pyrite oxidation at pH below 4 (Rawlings 2001). Fe- oxidizing chemolithotrophs gain energy from the oxidation of acidic ferrous Fe, although S is also used as an alternative electron donor. Sulfate reducers are generally found in the reducing conditions of anoxic environments. The sulfate-reducing bacteria play a vital role in pyrite formation under ASS. Dissimilatory sulfate reducers derive a large proportion of their metabolic energy from the reduction of sulfur and utilization of sulfur compounds as electron acceptors. Sulfate is reduced to sulfite in an eight-electron transfer reaction. In this process, a fixed carbon substrate such as ethanol or lactate is oxidized, and hydrogen sulfide gas (H_2S) is produced. The enzyme pathway responsible for the reduction of sulfur is known as the dissimilatory sulfur reductase system. Sequence analysis revealed the unique bacterial community assemblage present in the acid sulfate soil environment. A number of novel bacterial genera and species belonging to phyla *Proteobacteria*, *Actinobacteria*, *Bacteroidetes*, *Thermomicrobia*, *Verrucomicrobia*, *Firmicutes*, *Acidobacteria*, *Spirochaetes*, *Planctomycetes*, *Chloroflexi*, *Chlamydiae*, *Nitrospira*, *Dictyoglomi*, *Cyanobacteria*, *Deferribacteres*, *Gemmatimonadetes*, *Deinococcus-Thermus*, and *Fusobacteria* and the candidate phyla OP11 and OP10 were identified in the soil profile of a typical coastal acid sulfate soil under sugarcane cultivation. Analysis of the archaeal community composition through cloning-sequencing revealed the primary functions of these organisms in ASS environments were the production of methane and oxidation and reduction reactions of the sulfur cycle (Brunner et al. 2008).

4.7.3 Acid Mine Drainage and Coal Mine Spoils

Colliery spoils of all types contain some sulfur (S). Surface strip mine spoil having a pH <2 was found to contain 3–5% sulfur, which, in decreasing order of importance, was made up of (1) inorganic sulfidic S, (2) water-soluble S, (3) dilute acid-extractable S, (4) reduced S, (5) elemental S, and (6) anion-exchangeable S. Organic S contributed a minor fraction of the overall S content, although organic S may be present in some coal spoils (Harrison 1978). It is not surprising in view of the reduced inorganic S present that S oxidation occurs in colliery spoils. Brock (1978), using the $^{14}CO_2$ technique, showed that chemoautotrophic bacteria were found on the surface of pyritic materials associated with coal, but not below 10 cm depth. They also isolated large populations of heterotrophic bacteria and fungi, notably *Aureobasidium pullulans*, which is interesting considering that this fungus can oxidize sulfur in vitro (Williams and Cloete 2008). Populations of Fe- and S-oxidizing bacteria were also isolated from spoil in southeastern Montana, viz., waste coal (acidic pyrite-rich) and oxidized alkaline materials. S oxidation is generally hampered and decreased during summer and dry months. Not all forms of pyrite encountered in these spoils are subjected for bacterial oxidation. However, large crystals, for example, appeared to present too little surface area for rapid bacterial action. Application of SO_2 to calcareous spoils might improve their quality because *Thiobacillus* oxidize SO_2 to H_2SO_4, thereby lowering the excess alkalinity of the spoils and converting Na-saturated clay to Ca-saturated clay. Lack of sulfur oxidation due to dry spoils is unlikely to be a problem in areas with adequate rainfall. Problems relating to acid drainage from mines are often stressed in the literature (Kleinmann and Crerar 1979). In this respect, the activity of *T. ferrooxidans* has been emphasized. It is worth recalling that *T. ferrooxidans* can oxidize S^0 and $S_2O_3^{2-}$ with the formation of polythionates. For every mole of S oxidized, 180 mol of ferrous ion is oxidized but does not occur simultaneously. Harrison (1978) studied the microbial succession in an artificial coal spoil and showed that heterotrophic bacteria are an important component of the ecology of these habitats. Choline-SO_4^{-}

utilizing bacteria accounted for 1% of the population. Harrison (1978) suggested that organic sulfur present in coal may first be attacked by heterotrophs and the sulfur released may undergo further oxidation by *Thiobacillus*, particularly *T. ferrooxidans*.

4.7.4 Coastal Sand Dune

Coastal sand dunes are edaphic deserts and usually show nutrient deficiency for plant growth. The plant grown under this condition requires S from SO_4^{2-} deposited in sea spray (Skiba and Wainwright 1984). Coastal dunes tend to be S deficient due to the leaching out of $S0_4^{2-}$ rapidly. In these environments, elemental sulfur might be profitably used for increasing the amount of available sulfur in these environments. It was observed that S^0 was oxidized in sand and soil samples taken at various points along the dune succession, in which intermediates are formed in the form of SO_3 and $S_4O_2^-$. The S oxidation rate is generally enhanced by increasing content of C and N, decreasing in soil pH and vegetation cover. These sands tended to resist the acidification produced as S^0 was oxidized because of their high $CaCO_3$ contents; they might therefore be useful as sinks for waste gaseous S. The most occurring microbes of these ecosystems are *Salicornia* sp., *Puccinellia distans*, *Microcoleus chthonoplastes*, *Lyngbya aestuarii*, and *Leptolyngbya* sp. (Skiba and Wainwright 1984).

4.7.5 Hot Acid Soil

Although hot acid soils occur infrequently, they do provide an interesting habitat for the growth of heterotrophic microorganisms (Brock 1978). Solfatara is found in areas like Yellowstone Park. These are defined as areas where elemental S is precipitating out as a result of the oxidation of H_2S which are raised with steam from within the earth to the surface. They occupy hillsides, plateaus, small ravines, and shallow holes, and here springs are absent, but sulfur-rich soils at various temperatures are found, ranging in temperature from the mid-20 to the mid-30 °C range on the surface to about 75–90 °C at 20 cm depth. High concentrations of SO_2 are present (up to 152 mg g^{-1}), as are high levels of $S0_4^{2-}$ (4 mg g^{-1}), and pH values are as low as 0.7. *Thiobacillus* and *Sulfolobus* are present in these soils at the lower and higher temperatures (70 °C), respectively, and only overlap at 55 °C (Brock 1978).

4.7.6 Hot Spring

Hot springs are sites that release warm groundwater. The main possible reasons of high temperature in hot spring water are geothermal energy, exothermic reactions, and fission in radioactive elements (Mahala et al. 2013). Hot spring water usually have various kinds of minerals such as sulfates, carbonates, alkali, alkaline metals, and trace elements (Reddy et al. 2013); therefore, this is considered to have medicinal properties. Besides this, it also contains gasses like H_2S, CO_2, and low amount of O_2 (Mahala et al. 2013), and these gasses may be responsible for the sulfurous odor in hot spring water. Indian hot springs generally have moderate temperature (42–58 °C), moderate salinity, and near-neutral pH, whereas hot springs in other countries like the Philippines, China, and Malaysia have high temperature (50–110 °C), low to high salinities, and acidic or alkaline pH. Due to differences in these parameters (temperature, pH, and salinity), significantly dissimilar microbial phyla had been observed across tropical hot springs (Wang et al. 2013).

Moreover, in the hot spring environments, the important decomposers of organic matter under anoxic conditions are sulfate-reducing proteobacteria. Colorless sulfur bacteria can be isolated from sulfidic springs ranging from cold to mesophilic and geothermal hot sulfur springs. *Thiobacillus*, *Thiomonas*, *Beggiatoa*, and *Thiothrix* cells have been observed in the sulfidic springs of Frasassi cave system. *Beggiatoa* populations normally flourish in microaerophilic environment than *Thiothrix* (Macalady et al. 2006). *Themothrix azorensis* an obligately chemolithoautotrophic, thermophile growing in temperature range of 63–86 °C, was isolated from a hot spring (Odintsova et al. 1996). *Thiomicrospira psychrophila*, *Thiobacillus*, and *Halothiobacillus* sp. strain RA13 were reported from Gypsum Hill and Colour Peak sulfur springs; *Thiomicrospira* was dominant in sediment microbial communities as indicated by DNA-based analysis (Perreault et al. 2007). It was observed that few novel microbial species such as *Thiomonas bhubaneswarensis*, *Chelatococcus sambhunathii*, *Comamonas thiooxydans*, and *Gulbenkiania indica* were isolated from the four tropical hot springs of Odisha (India), namely, Taptapani, Tarabalo, Atri, and Athmallik (Jyoti et al. 2010; Narayan et al. 2016). Some of the thermotolerant plant growth-promoting fungi were also isolated from hot springs of Odisha and registered in National Fungal Culture Collection of India (NFCCI), Pune, by Kumar and Dangar (2014). Genus *Sulfolobus* was discovered from hot springs and is a thermophilic, acidophilic, facultative autotroph. *Thermothrix thioparus*, a neutrophilic thermophile, capable of depositing sulfur extracellularly and oxidizing sulfur compounds anaerobically using nitrate, was recovered from a New Mexico hot spring, whereas a sulfur oxidizer bacterium, *Sulfurihydrogenibium yellowstonense*, extremely thermophilic, facultatively heterotrophic, was isolated from Yellowstone National Park. Occurrence of *Sulfurovum-like* spp. with *Thiothrix* and *Thiofaba* spp. was reported from sulfur springs in the USA. Sulfide concentration in the environment also affects diversity of colorless sulfur bacteria. Based on molecular diversity analysis, *Chloroflexus* and *Aquificales* were found dominant in the low-sulfide spring and high-sulfide spring, respectively, at the same temperature (Skirnisdottir et al. 2000).

4.7.7 Marine Water and Sediments

In marine habitats, the initial step of S cycle is the oxidation of hydrogen sulfide (H_2S). However, microbial role of sulfur oxidation under these habitats especially marine sediments is largely unknown, with exception of certain mat-forming and filamentous bacteria (Jørgensen 1982). In marine system, the sulfur-oxidizing prokaryotes generally are able to oxidize H_2S present in sulfidic intertidal sediments which are produced by sulfate-reducing microbes after utilizing oxidized S compounds as substrate (Jørgensen 1982). Other researchers indicated that in freshwater ecosystem (flooded rice field), the sulfur concentration is slightly lower than the marine ecosystem.

4.7.8 Peatland Soil

Peatland ecosystem is formed due to long-term incremental increase of global warming, less precipitation, and atmospheric deposition of reactive nitrogen and sulfur compounds, accompanied by unforeseeable changes in the carbon balance (Dise 2009). It is estimated that peatlands can emit methane which constitutes 10–20% of the total global methane emission (Wuebbles and Hayhoe 2002) and increase global atmospheric sulfur pollution and acid precipitation (Gauci et al. 2004). In peatland soil, anoxic recycling of reduced sulfur compounds accompanied by high sulfate reduction rates resulted in the formation of "thiosulfate shunt" (Blodau et al. 2007). Some of the important factors responsible for this process in peatland ecosystem are vegetation type, drought, and alternating periods of precipitation (Wind and Conrad 1997; Paul et al. 2006; Reiche et al. 2009; Deppe et al. 2010). One representative model (fen system) for peatland system is located at forested Lehstenbach catchment (Bavaria, Germany) which gives the significance of dissimilatory sulfate reduction by microbes in this system (Klemm and Lange 1999; Alewell et al. 2000).

Relatively lower abundance of *Desulfosporosinus* species (only 0.006% of the total bacterial and archaeal 16S rRNA genes) were encountered under peatland system; however substantial capacity of sulfate reduction was catalyzed by them only. On the other hand, a large portion of sulfate reduction under in situ still remains unsolved (Pester et al. 2010). Mostly in peatland, microbial-mediated dissimilatory (bi) sulfite reductase (*dsrAB gene*) is operated that utilizes sulfite or sulfate anaerobically; that is why these genes act as suitable markers to assess molecular diversity studies in peatland (Dhillon et al. 2003; Kjeldsen et al. 2007). *Desulfomonile* and *Syntrophobacter* were occasionally detected by *dsrABFGA* analysis and generally present in lower soil layer than in the deeper soil layers (Steger et al. 2011). In peatlands, usually the position of the water table marks the transition between the oxic and anoxic zones. Novel *dsrAB*-carrying microorganisms are widespread in wetlands, and *dsrB* DGGE bands and a *dsrAB* clone library revealed that these were broadly distributed among different bogs and fens and related to *Syntrophobacter wolinii* (Pester et al. 2010). However, the relatively high abundance of unique microflora are yet to be discovered under model peatland ecosystem which would be desirable future research to better understand the nutrient cycle including S cycle under this system (Stepanauskas and Sieracki 2007; Wagner 2009; Xie et al. 2005).

4.8 Conclusion and Future Prospects

It has been established that most of the sulfur compounds utilized by plants for their growth is derived from soil organosulfur pool and the mobilization and assimilation of sulfur by plants are mediated by the soil microbial community. The main drivers of sulfur biogeochemical process in different ecosystems are bacteria and archaea. So far, very limited studies have been conducted to prove beneficial effect of inoculation with sulfur-oxidizing bacteria (SOB), and also no commercial product is available elsewhere on SOB-based bioformulations. Recently, researchers attempted to use granular form of elemental S (ES) (Zhao et al. 2017a) and ES-Zn (Mattiello et al. 2017) fertilizers with the help of S-oxidizing microorganisms, and they further indicated that this form of S is slower to oxidize than powdered elemental S mixed through soil (Zhao et al. 2017a). They also suggested that ES oxidation was not affected by short-term changes in bacterial abundance and community composition by temporary increases in soil acidity or ionic strength (Zhao et al. 2017b). Some researchers also revealed for the first time that besides common SOB, two other groups of bacteria (*Comamonadaceae* and *Rhodococcus*) may also play a specialized role in sulfonate cycling in the soil (Schmalenberger et al. 2009). In addition, mycorrhizal fungi and protozoa in association with bacteria are also important in providing sulfur to plants. Till date, researchers have made considerable advances for understanding how soil organosulfur is converted to plant-available sulfur as well as their regulating mechanism of this process. However, further in-depth investigations are required to understand S transformation process under different habitats through integrated molecular ecology approach as sulfur cycling becomes an important component in anthropogenic ecosystem environment.

References

Agrawal A, Lal B (2009) Rapid detection and quantification of bisulfite reductase genes in oil field samples using real-time PCR. FEMS Microbiol Ecol 69:301–312

Alewell C, Manderscheid B, Meesenburg H, Bittersohl J (2000) Is acidification still an ecological threat. Nature 407:856–858

Anandham JJ (1991) Sulfur-oxidizing bacteria as plant growth promoting rhizobacteria for canola. Can J Microbiol 7:521–529

Anandham R, Gandhi PI, Kwon SW, Sa TM, Kim YK, Jee HJ (2009) Mixotrophic metabolism in *Burkholderia kururiensis* subsp. *thiooxydans* subsp. *nov*., a facultative chemolithoautotrophic thiosulfate oxidizing bacterium isolated from rhizosphere soil and proposal for classification of the type strain of *Burkholderia kururiensis* as *Burkholderia kururiensis* subsp. *kururiensis* subsp. *nov*. Arch Microbiol 191:885–894

Appia-Ayme C, Little PJ, Matsumoto Y, Leech AP, Berks BC (2001) Cytochrome complex essential for photosynthetic oxidation of both thiosulfate and sulfide in *Rhodovulum sulfidophilum*. J Bacteriol 183:6107–6118

Badhai J, Ghosh TS, Das SK (2014) Taxonomic and functional characteristics of microbial communities and their correlation with physicochemical properties of four geothermal springs in Odisha, India. Front Microbiol 6:1166

Bamford VA, Bruno S, Rasmussen T, Appia-Ayme C, Cheesman MR, Berks BC, Hemmings AM (2002) Structural basis for the oxidation of thiosulfate by a sulfur cycle enzyme. EMBO J 21:5599–5610

Bardischewsky F, Quentmeier A, Rother D, Hellwig P, Kostka S, Friedrich CG (2005) Sulfur dehydrogenase of *Paracoccus pantotrophus*: the heme-2 domain of the molybdoprotein cytochrome c complex is dispensable for catalytic activity. Biochemistry 44:7024–7034

Bettany JR, Stewart JWB, Halstead EH (1973) Sulfur fractions and carbon, nitrogen and sulfur relationships in grassland, forest and associated transitional soils. Soil Sci Soc Am Proc 37:915–918

Blodau C, Mayer B, Peiffer S, Moore TR (2007) Support for an anaerobic sulfur cycle in two Canadian peatland soils. J Geophys Res 112:G02004

Brock TD (1978) Thermophilic micro-organisms and life at high temperatures. Springer, New York, pp 1–465

Brune DC (1989) Sulfur oxidation by phototrophic bacteria. Biochim Biophys Acta 975:189–221

Brune DC, Blankenship RE, Madigan MT, Bauer CE (1995) Sulfur compounds as photosynthetic electron donors Anoxygenic photosynthetic bacteria. Kluwer Academic Publishers, Dordrecht, pp 847–870

Brunner B, JY Y, Mielke RE, MacAskill JA, Madzunkov S, McGenity TJ, Coleman M (2008) Different isotope and chemical patterns of pyrite oxidation related to lag and exponential growth phases of *Acidithiobacillus ferrooxidans* reveal a microbial growth strategy. Earth Planet Sci Lett 270:63–72

Burke ME, Gorham E, Pratt DC (1974) Distribution of purple photosynthetic bacteria in wetland and woodland habitats of central and northern Minnesota. J Bacteriol 117:826–833

Chapman SJ (1990) *Thiobacillus* populations in some agricultural soils. Soil Biol Biochem 22:479–482

Dahl C, Friedrich RW (2008) Inorganic sulfur compounds as electron donors in purple sulfur bacteria, vol 168. Institut für Mikrobiologie Biotechnologie, Bonn, p D-53115

Dahl C, Trüper HG (1994) Enzymes of dissimilatory sulfide oxidation in phototrophic bacteria. Methods Enzymol 243:400–421

Dahl C, Engels S, Pott-Sperling AS, Schulte A, Sander J, Lübbe Y, Deuster O, Brune DC (2005) Novel genes of the dsr gene cluster and evidence for close interaction of Dsr proteins during sulfur oxidation in the phototrophic sulfur bacterium *Allochromatium vinosum*. J Bacteriol 187:1392–1404

Dent DL (1986) Acid sulphate soils: a baseline for research and development. ILRI Publications, Wageningen, p 39

Deppe M, McKnight DM, Blodau C (2010) Effects of short-term drying and irrigation on electron flow in mesocosms of a northern bog and an alpine fen. Environ Sci Technol 44:80–86

de Zwart JMM, Nelisse PN, Kuenen JG (1996) Isolation and characterization of *Methylophaga sulfidoVorans, sp.* nov.: an obligately methylotrophic, aerobic, dimethyl sulfide oxidizing bacterium from a microbial mat. FEMS Microbiol Ecol 20:261–270

Dhillon A, Teske A, Dillon J, Stahl DA, Sogin ML (2003) Molecular characterization of sulfate-reducing bacteria in the Guaymas Basin. Appl Environ Microbiol 69:2765–2772

Dise NB (2009) Peatland response to global change. Science 326:810–811

Fischer U (1989) Enzymatic steps in dissimilatory sulfur metabolism by whole cells of anoxyphotobacteria. In: Saltzman E, Cooper W (eds) Biogenic Sulfur in the Environment. American Chemical Society, Washington, DC, pp 262–279

Freney JR, Jacq VA, Baldensperger J (1982) Microbiology of tropical soils and plant productivity. Martinus Nijhoff Publishers, The Hague, pp 271–317

Friedrich CG, Mitrenga G (1981) Oxidation of thiosulfate by *Paracoccus denitrificans* and other hydrogen bacteria. FEMS Microbiol Lett 10:209–212

Friedrich CG, Rother D, Bardischewsky F, Quentmeier A, Fischer J (2001) Oxidation of inorganic sulfur compounds by bacteria: emergence of a common mechanism? Appl Environ Microbiol 67:2873–2882

Friedrich CG, Bardischewsky F, Rother D, Quentmeier A, Fischer J (2005) Prokaryotic sulfur oxidation. Curr Opin Microbiol 8:253–259

Frigaard NU, Bryant DA (2008) Genomic insights into the sulfur metabolism of phototrophic green sulfur bacteria. Springer, Dordrecht, pp 337–355

Gauci V et al (2004) Sulfur pollution suppression of the wetland methane source in the 20th and 21st centuries. Proc Natl Acad Sci 101:12583–12587

Gentile R, Vanlauwe B, Chivenge P, Six J (2011) Trade-offs between the short-and long-term effects of residue quality on soil C and N dynamics. Plant Soil 338:159–169

Germida JJ, Wainwright M, Gupta VV (1992) Biochemistry of sulfur cycling in soil. Soil Biochem 7:1–53

Ghosh W, Dam B (2009) Biochemistry and molecular biology of lithotrophic sulfur oxidation by taxonomically and ecologically diverse bacteria and archaea. FEMS Microbiol Rev 33:999–1043

Grayston SJ, Wainwright M (1988) Sulphur oxidation by soil fungi including some species of mycorrhizae and wood-rotting basidiomycetes. FEMS Microbiol Ecol 4:1–8

Grabarczyk DB, Chappell PE, Johnson S, Stelzl LS, Lea SM, Berks BC (2015) Structural basis for specificity and promiscuity in a carrier protein/enzyme system from the sulfur cycle. Proc Natl Acad Sci 112:7166–7175

Grayston SJ, Nevell W, Wainwright M (1986) Sulphur oxidation by fungi. Trans Br Mycol Soc 87:193–198

Harrison AP (1978) Microbial succession and mineral leaching in an artificial coal spoil. Appl Environ Microbiol 36:861–869

He JZ, Liu XZ, Zheng Y, Shen JP, Zhang LM (2010) Dynamics of sulfate reduction and sulfate-reducing prokaryotes in anaerobic paddy soil amended with rice straw. Biol Fertil Soils 46:283–291

Hensen D, Sperling D, Trüper HG, Brune DC, Dahl C (2006) Thiosulfate oxidation in the phototrophic sulfur bacterium *Allochromatium vinosum*. Mol Microbiol 62:794–810

Hipp WM, Pott AS, Thum-Schmitz N, Faath I, Dahl C, Trüper HG (1997) Towards the phylogeny of APS reductases and sirohaem sulphite reductases in sulfate- reducing and sulfur-oxidizing prokaryotes. Microbiology 143:2891–2902

Imhoff JF (2003) Phylogenetic taxonomy of the family *Chlorobiaceae* on the basis of 16S rRNA and fmo (Fenna-Matthews–Olson protein) gene sequences. Int J Syst Evol Microbiol 53:941–951

Imhoff JF, Hiraishi A (2005) Aerobic bacteria containing bacteriochlorophyll and belonging to the *Alphaproteobacteria*. In: Brenner DJ, Krieg NR, Staley JT, Garrity GM (eds) Bergey's manual of systematic bacteriology, vol 2. Springer, New York, p 135

Janosch C, Remonsellez F, Sand W, Vera M (2015) Sulfur Oxygenase Reductase (Sor) in the moderately Thermoacidophilic leaching bacteria: studies in *Sulfobacillus thermosulfidooxidans* and *Acidithiobacillus caldus*. In: Amils R, Toril EG (eds) Microorganisms 3:707–724

Jensen LS, Salo T, Palmason F, Breland TA, Henriksen TM, Stenberg B, Pedersen A, Lundström C, Esala M (2005) Influence of biochemical quality on C and N mineralisation from a broad variety of plant materials in soil. Plant Soil 273:307–326

Jørgensen BB (1982) Mineralization of organic matter in the sea bed – the role of sulphate reduction. Nature 296:643–645

Jørgensen BB, Nelson DC (2004) Sulfide oxidation in marine sediments: geochemistry meets microbiology. Geol Soc Am Spec Pap 379:63–81

Joshi MM, Hollis JP (1976) Rapid enrichment of *Beggiatoa* from soil. J Appl Bacteriol 40:223–224

Jyoti V, Narayan KD, Das SK (2010) *Gulbenkiania indica* sp. nov, isolated from a sulfur spring. Int J Syst Evol Microbiol 60:1052–1055

Kappler U, Dahl C (2001) Enzymology and molecular biology of prokaryotic sulfite oxidation. FEMS Microbiol Lett 203:1–9

Kelly DP, Shergill JK, WP L, Wood AP (1997) Oxidative metabolism of inorganic sulfur compounds by bacteria. Antonie Van Leeuwenhoek 71:95–107

Kertesz MA, Mirleau P (2004) The role of soil microbes in plant sulfur nutrition. J Exp Bot 55:1939–1945

Khanna S, Nicholas DJD (1982) Utilization of tetrathionate and 35S-labelled thiosulphate by washed cells of *Chlorobium vibrioforme* of sp *Thiosulfatophilum*. J Gen Microbiol 128:1027–1034

Kjeldsen KU et al (2007) Diversity of sulfate-reducing bacteria from an extreme hypersaline sediment, Great Salt Lake (Utah). FEMS Microbiol Ecol 60:287–298

Kleinmann RL, Crerar DA (1979) *Thiobacillus ferrooxidans* and the formation of acidity in simulated coal mine environments. Geomicrobiol J 1:373–388

Klemm O, Lange H (1999) Trends of air pollution in the Fichtelgebirge Mountains, Bavaria. Environ Sci Pollut Res Int 6:193–199

Klotz MG, Bryant DA, Hanson TE (2011) The microbial sulfur cycle. Front Microbiol 2:1–2

Knights JS, Zhao FJ, McGrath SP, Magan N (2001) Long-term effects of land use and fertiliser treatments on sulphur transformations in soils from the Broadbalk experiment. Soil Biol Biochem 33:1797–1804

Krishnani KK, Kathiravan V, Natarajan M, Kailasam M, Pillai SM (2010) Diversity of sulfur-oxidizing bacteria in greenwater system of coastal aquaculture. Appl Biochem Biotechnol 162:1225–1237

Kuenen JG, Tuovinen DH (1981) The genera *Thiobacillus* and *Thiomicrospira*. In: Starr MP et al (eds) The prokaryotes, a handbook on habitats, isolation and identification of bacteria. Springer, New York, pp 1023–1036

Kumar U, Dangar TK (2014) Thermo-tolerant plant-growth promoting fungi (PGPF) from hot springs of Odisha. CRRI Newslett 35:8–9. http://www.crri.nic.in/CRRI_newsletter/crnl_aprjune_2014_web.pdf

Kumar U, Berliner J, Adak T, Rath PC, Dey A, Pokhare SS, Jambhulkar NN, Panneerselvam P, Kumar A, Mohapatra SD (2017) Non-target effect of continuous application of chlorpyrifos on soil microbes, nematodes and its persistence under sub-humid tropical rice-rice cropping system. Ecotoxicol Environ Saf 135:225–235

Langenhoff R (1986) Distribution, mapping, classification and use of acid sulphate soils in the tropics, a literature study. Soil Survey Institute, Wageningen, p 133

Larsen H (1952) On the culture and general physiology of the green sulfur bacteria. J Bacteriol 64:187–196

Larsen Ø, Lien T, Birkeland NK (2001) A novel organization of the dissimilatory sulfite reductase operon of *Thermodesulforhabdus norvegica* verified by RT-PCR. FEMS Microbiol Lett 203:81–85

Lawrence JR, Gupta VVSR, Germida JJ (1988) Impact of elemental sulfur fertilization on agricultural soils II Effects on sulfur oxidizing populations and oxidation rates. Can J Soil Sci 68:475–483

Lübbe YJ, Youn H-S, Timkovich R, Dahl C (2006) Siro (haem) amide in *Allochromatium vinosum* and relevance of DsrL and DsrN, a homolog of cobyrinic acid a, c diamide synthase for sulfur oxidation. FEMS Microbiol Lett 261:194–202

Lucheta AR, Lambais MR (2012) Sulfur in agriculture. Revista Brasileira de Ciência do Solo 36:1369–1379

Macalady JL, Lyon EH, Koffman B, Albertson LK, Meyer K, Galdenzi S, Mariani S (2006) Dominant microbial populations in limestone-corroding stream biofilms, Frasassi cave system, Italy. Appl Environ Microbiol 72:5596–5609

Mahala SC, Singh P, Das M, Acharya S (2013) Genesis of thermal springs of Odisha, India. Int J Earth Sci Eng 5:1572–1577

Mander GJ, Pierik AJ, Huber H, Hedderich R (2004) Two distinct heterodisulfide reductase-like enzymes in the sulfate-reducing archaeon *Archaeoglobus profundus*. Eur J Biochem 271:1106–1116

Mathew EK, Panda RK, Nair M (2001) Influence of subsurface drainage on crop production and soil quality in a low-lying acid sulphate soil. Agric Water Manag 47:191–209

Mattiello EM, da Silva RC, Degryse F, Baird R, Gupta VV, McLaughlin ML (2017) Sulfur and zinc availability from co-granulated Zn-enriched elemental sulfur fertilizers. J Agric Food Chem 65:1108–1115

McLaren RG, Keer JI, Swift RS (1985) Sulfur transformations in soils using S-35 labeling. Soil Biol Biochem 17:73–79

Meyer B, Kuever J (2007) Phylogeny of the alpha and beta subunits of the dissimilatory adenosine-5′-phosphosulfate (APS) reductase from sulfate reducing prokaryotes – origin and evolution of the dissimilatory sulfate-reduction pathway. Microbiology 153:2026–2044

Mouraret M, Baldensperger J (1977) Use of membrane filters for the enumeration of autotrophic *Thiobacilli*. Microb Ecol 3:345–358

Narayan KD, Sabat SC, Das SK (2016) Mechanism of electron transport during thiosulfate oxidation in an obligately mixotrophic bacterium *Thiomonas bhubaneswarensis* strain S10 (DSM 18181T). Appl Microbiol Biotechnol 10:1–4

Nelson DC, Fisher CR (1995) Chemoautotrophic and methanotrophic endosymbiotic bacteria at deep-sea vents and seeps, in the microbiology of deep-sea hydrothermal vents. CRC Press, Boca Raton, pp 125–167

Niknahad-Gharmakher H, Piutti S, Machet JM, Benizri E, Recous S (2012) Mineralization-immobilization of sulphur in a soil during decomposition of plant residues of varied chemical composition and S content. Plant Soil 360:391–404

Odintsova EV, Jannasch HW, Mamone JA, Langworthy TA (1996) *Thermothrix azorensis* sp. *nov*., an obligately chemolithoautotrophic, sulfur-oxidizing, thermophilic bacterium. Int J Syst Evol Microbiol 46:422–428

Paul S, Kusel K, Alewell C (2006) Reduction processes in forest wetlands: tracking down heterogeneity of source/sink functions with a combination of methods. Soil Biol Biochem 38:1028–1039

Pelletier N, Leroy G, Guiral M, Giudici-Orticoni MT, Aubert C (2008) First characterisation of the active oligomer form of sulfur oxygenase reductase from the bacterium *Aquifex aeolicus*. Extremophiles 12:205–215

Perreault NN, Andersen DT, Pollard WH, Greer CW, Whyte LG (2007) Characterization of the prokaryotic diversity in cold saline perennial springs of the Canadian high Arctic. Appl Environ Microbiol 73:1532–1543

Pester M, Bittner N, Deevong P, Wagner M, Loy A (2010) A 'rare biosphere' microorganism contributes to sulfate reduction in a peatland. ISME J 4:1591–1602

Petri R, Podgorsek L, Imhoff JF (2001) Phylogeny and distribution of the *soxB* gene among thiosulfate oxidizing bacteria. FEMS Microbiol Lett 197:171–178

Plumb JJ, Haddad CM, Gibson JA, Franzmann PD (2007) *Acidianus sulfidivorans* sp. *nov*., an extremely acidophilic, thermophilic archaeon isolated from a solfatara on Lihir Island, Papua New Guinea, and emendation of the genus description. Int J Syst Evol Microbiol 57:1418–1423

Rawlings DE (2001) The molecular genetics of *Thiobacillus ferrooxidans* and other mesophilic, acidophilic, chemolithotrophic, iron- or sulfur-oxidizing bacteria. Hydrometallurgy 59:187–201

Reddy DV, Nagbhusanam P, Ramesh G (2013) Turnover time of rural and Rajvadi hot spring waters, Maharastra, India. Curr Sci 104:1419–1424

Reiche M, Hädrich A, Lischeid G, Küsel K (2009) Impact of manipulated drought and heavy rainfall events on peat mineralization processes and source-sink functions of an acidic fen. J Geophys Res 114:G02021

Roberts TL, Bettany JR (1985) The influence of topography on the nature and distribution of soil sulfur across a narrow environmental gradient. Can J Soil Sci 65:419–434

Sahoo K, Dhal NK (2009) Dhal Potential microbial diversity in mangrove ecosystems: a review. Indian J Mar Sci 38:249–256

Schauder R, Kröger A (1993) Bacterial sulphur respiration. Arch Microbiol 159:491–497

Schedel M, Vanselow M, Truper HG (1979) Siroheme sulfite reductase from *Chromatium vinosum* purification and investigation of some of its molecular and catalytic properties. Arch Microbiol 121:29–36

Schmalenberger A, Hodge S, Hawkesford MJ, Kertesz MA (2009) Sulfonate desulfurization in *Rhodococcus* from wheat rhizosphere communities. FEMS Microbiol Ecol 67:140–150

Skiba U, Wainwright M (1984) Oxidation of elemental-S in coastal-dune sands and soils. Plant Soil 77:87–95

Skirnisdottir S, Hreggvidsson GO, Hjörleifsdottir S, Marteinsson VT, Petursdottir SK, Holst O, Kristjansson JK (2000) Influence of sulfide and temperature on species composition and community structure of hot spring microbial mats. Appl Environ Microbiol 66:2835–2841

Smith AJ, Lascelles J (1966) Thiosulphate metabolism and rhodanese in *Chromatium sp.* strain D. J Gen Microbiol 42:357–370

Sorokin DY, Tourova TP, Galinski EA, Muyzer G, Kuenen JG (2008) *Thiohalorhabdus denitrificans* gen. *nov.*, sp. *nov.*, an extremely halophilic, sulfur-oxidizing, deep-lineage gammaproteobacterium from hypersaline habitats. Int J Syst Evol Microbiol 58:2890–2897

Starkey RL (1934) The production of polythionates from thiosulfate by microörganisms. J Bacteriol 28:387

Steger D, Wentrup C, Braunegger C (2011) Microorganisms with novel dissimilatory (Bi)sulfite reductase genes are widespread and part of the core microbiota in low-sulfate peatlands. Appl Environ Microbiol 77:1231–1242

Steinmetz MA, Fischer U (1982) Cytochromes of the green sulfur bacterium *Chlorobium vibrioforme thiosulfatophilum*, purification, characterization and sulfur metabolism. Arch Microbiol 19:19–26

Stepanauskas R, Sieracki ME (2007) Matching phylogeny and metabolism in the uncultured marine bacteria, one cell at a time. Proc Natl Acad Sci U S A 104:9052–9057

Tabatabai MA (1984) Importance of sulfur in crop production. Biogeochemistry 1:45–62

Takakuwa S, Oae S, Okuyama T (1992) Biochemical aspects of microbial oxidation of inorganic sulfur compounds, in Organic Sulfur Chemistry, Biochemical Aspects. CRC Press, Boca Raton, pp 1–43

Tang Y, Pingitore F, Mukhopadhyay A, Phan R et al (2007) Pathway confirmation and flux analysis of cental metabolic pathways in *D. vulgaris*, Hildenborough, using gas chromatography, mass spectrometry and fourtier transform ion cyclotrone resonance mass spectroscopy. J Bacteriol 189:940–949

Then J, Trüper HG (1981) The role of thiosulfate in sulfur metabolism of *Rhodopseudomonas globiformis*. Arch Microbiol 130:143–146

Tourna M, Maclean P, Condron L, O'Callaghan M, Wakelin SA (2014) Links between sulphur oxidation and sulphur-oxidising bacteria abundance and diversity in soil microcosms based on functional gene analysis. FEMS Microbiol Ecol 88:538–549

Tourova TP, Kovaleva OL, Bumazhkin BK, Patutina EO, Kuznetsov BB, Bryantseva IA, Gorlenko VM, Sorokin DY (2011) Application of ribulose-1, 5-bisphosphate carboxylase/oxygenase genes as molecular markers for assessment of the diversity of autotrophic microbial communities inhabiting the upper sediment horizons of the saline and soda lakes of the Kulunda Steppe. Microbiology 80:812–825

Trüper HG, Fischer U (1982) Anaerobic oxidation of sulfur compounds as electron donors for bacterial photosynthesis. Philos Trans R Soc Lond Ser B Biol Sci 298:529–542

Trüper HG, Pfennig N (1966) Sulfur metabolism in Thiorhodaceae. III. Storage and turnover of thiosulphate sulfur in *Thiocapsa floridana* and *Chromatium species*. Int J Gen Mol Microbiol 32:261–276

Tuttle JH (1980) Organic carbon utilization by resting cells of thiosulfate-utilizing marine heterotrphs. Appl Environ Microbiol 40:516–521

Urich T, Coelho R, Kletzin A, Frazao C (2005) The sulfur oxygenase reductase from *Acidianus ambivalens* is an icosatetramer as shown by crystallization and Patterson analysis. Biochim Biophys Acta 1747:267–270

Vermeij P, Wietek C, Kahnert A, Wüest T, Kertesz MA (1999) Genetic organization of sulfurcontrolled aryl desulfonation in *Pseudomonas putida* S-313. Mol Microbiol 32:913–926

Wagner M (2009) Single-cell ecophysiology of microbes as revealed by Raman microspectroscopy or secondary ion mass spectrometry imaging. Annu Rev Microbiol 63:411–429

Wagner M, Roger AJ, Flax JL, Brusseau GA, Stahl DA (1998) Phylogeny of dissimilatory sulfite reductases supports an early origin of sulfate respiration. J Bacteriol 180:2975–2982

Wainwright M (1984) Sulfur oxidation in soils. Adv Agron 37:349–396

Wakai S, Kikumoto M, Kanao T, Kamimura K (2004) Involvement of sulfide: quinone oxidoreductase in sulfur oxidation of an acidophilic iron-oxidizing bacterium, *Acidithiobacillus ferrooxidans* NASF-1. Biosci Biotechnol Biochem 68:2519–2528

Wang S, Hou W, Dong H, Jiang H, Huang L, Wu G, Zhang C, Song Z, Zhang Y, Ren H, Zhang J (2013) Control of temperature on microbial community structure in hot springs of the Tibetan Plateau. PLoS One 8:e62901

White R, Engelen G (1997) Cellular automata as the basis of integrated dynamic regional modelling. Environ Plann B Plann Des 24:235–246

Williams CH (1972) Sulfur deficiency in Australia. Sulfur Inst J 8:5–8

Williams PJ, Cloete TE (2008) Microbial community study of the iron ore concentrate of the Sishen Iron Ore Mine, South Africa. World J Microbiol Biotechnol 24:2531–2538

Wind T, Conrad R (1997) Localization of sulfate reduction in planted and unplanted rice field soil. Biogeochemistry 37:253–278

Wodara C, Bardischewsky F, Friedrich CG (1997) Cloning and characterization of sulfite dehydrogenase, two c-type cytochromes, and a flavoprotein of *Paracoccus denitrificans* GB17: essential role of sulfite dehydrogenase in lithotrophic sulfur oxidation. J Bacteriol 179:5014–5023

Wu J, O'Donnell AG, Syers JK (1995) Influences of glucose, nitrogen and plant residues on the immobilization of sulphate-S in soil. Soil Biol Biochem 27:1363–1370

Wu QL, Zwart G, Schauer M, Kamst-van Agterveld MP, Hahn MW (2006) Bacterioplankton community composition along a salinity gradient of sixteen high-mountain lakes located on the Tibetan Plateau, China. Appl Environ Microbiol 72:5478–5485

Wuebbles DJ, Hayhoe K (2002) Atmospheric methane and global change. Earth Sci Rev 57:177–210

Xie C, Chen D, Li YQ (2005) Raman sorting and identification of single living micro-organisms with optical tweezers. Opt Lett 30:1800–1802

Yao H, Conrad R, Wassmann R, Neue HU (1999) Effect of soil characteristics on sequential reduction and methane production in sixteen rice paddy soils from China, the Philippines, and Italy. Biogeochemistry 47:269–295

Yousuf B, Kumar R, Mishra A, Jha B (2014) Unravelling the carbon and sulfur metabolism in coastal soil ecosystems using comparative cultivation independent genome-level characterisation of microbial communities. PLoS One 9:e107025

Zhao C, Gupta VV, Degryse F, McLaughlin MJ (2017a) Abundance and diversity of sulphur-oxidising bacteria and their role in oxidising elemental sulphur in cropping soils. Biol Fertil Soils 53:159

Zhao C, Gupta VV, Degryse F, McLaughlin MJ (2017b) Effects of pH and ionic strength on elemental sulphur oxidation in soil. Biol Fertil Soils 53:247

Microbial Cycling of Arsenic in the Aquifers of Bengal Delta Plains (BDP)

5

Devanita Ghosh and Punyasloke Bhadury

Abstract

The element arsenic is abundant on earth's crust. Solubilization due to physicochemical or biologically mediated processes can lead to elevated levels of arsenic in the aquatic environments. In the Bengal Delta Plains (BDP), arsenic mobilization, in particular in aquifers and agricultural lands, has resulted in serious health manifestations among human populations residing in India and Bangladesh. Moreover, the organic matter composition has been shown to be the key component controlling arsenic fluxes in groundwater of BDP region. Microbes have the capability to alter As fluxes and thus can form the basis of cost-effective bioremediation technologies for As-free drinking water. In this chapter emphasis has been laid on the distribution of As and its fluxes across different ecosystems. The fluxes that are controlled by microbial metabolic pathways, which in turn depend on bioavailability and properties of organic matter in the environment, have been highlighted in this chapter.

Keywords

Arsenic · Cycling · Bengal Delta Plains · Fluxes · Metabolic pathways

D. Ghosh
Laboratory of Biogeochemistry, Centre for Earth Sciences, Indian Institute of Science, Bangalore, India

P. Bhadury (✉)
Integrative Taxonomy and Microbial Ecology Research Group, Department of Biological Sciences, Indian Institute of Science Education and Research Kolkata, Mohanpur, West Bengal, India
e-mail: pbhadury@iiserkol.ac.in

T. K. Adhya et al. (eds.), *Advances in Soil Microbiology: Recent Trends and Future Prospects*, Microorganisms for Sustainability 3,
https://doi.org/10.1007/978-981-10-6178-3_5

5.1 Introduction

The element arsenic (As) is one of the most well-known toxic compounds on earth, having atomic number 33, and mainly found in various mineral forms in conjunction with sulfur and other metals. The element was first documented in 1250 AD by the German alchemist Albertus Magnus (Emsley 2001). However, As in crystalline form is very rare, comprising around 0.00005% of earth's crust (Gulledge and O'Connor 1973), and its average abundance in sedimentary and igneous rocks is relatively less (~2 mg/kg) (Kabata-Pendias and Pendias 1984). However, fine-grained argillaceous sediments and phosphorites may have higher As abundance in comparison to igneous rock (Mandal and Suzuki 2002). Physicochemical conditions can affect solubilization and/or microbial mobilization and subsequent bioavailability of As in any environment (Smedley and Kinniburgh 2002). Some As-derived compounds get easily dissolved under varying physicochemical and microbiological environments (Fig. 5.1). Such dissolution may lead to elevated levels of As in natural water bodies.

Human populations depending on such water for drinking, agriculture, and other potable usage become prone to several critical health manifestations including various forms of cancer. The most alarming case is observed in India and

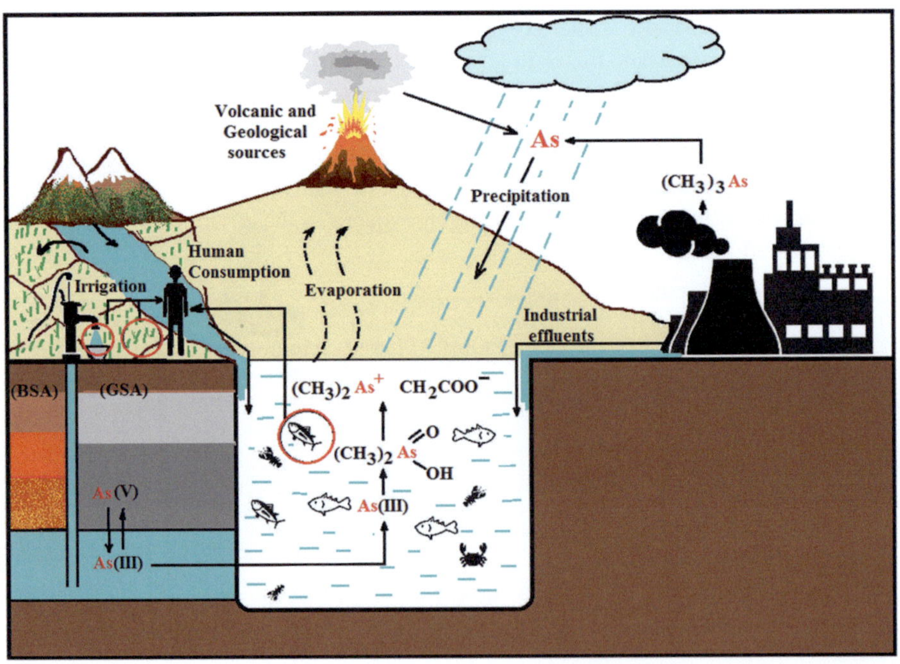

Fig. 5.1 Global arsenic cycling representing different sources such as the volcanoes and thermal springs, industrial effluents, and geogenic origin. Arsenic from these sources is ultimately transported into the human body through various trophic levels in the food chain

Bangladesh where dissolution of As into groundwater has exposed more than 70 million people to various health risks directly linked to As. Such exposure of naturally occurring arsenic to human population can pass through food chain like vegetables (Roychowdhury et al. 2002; Das et al. 2004), rice (Warren et al. 2003), fish (Das et al. 2004; Shah et al. 2009), and milk (Roychowdhury et al. 2002). The contamination of crops such as paddy in the large fertile plains of Bengal with As is known to have enormous socioeconomic consequences (Debnath and Bhadury 2016).

The Bengal Delta Plains (BDP) are formed on the riverine deposits of three main rivers, namely, Ganges, Brahmaputra, and Meghna, which also contain large arsenopyrite deposit. The organic matter composition of the region has been shown to be the key component controlling arsenic fluxes in groundwater (Lawson et al. 2013; Ghosh et al. 2015a). However, anoxic condition also affects the dissolution of As minerals into water bodies thereby increasing its bioavailability. Due to the considerable threat of severe health manifestations due to As exposure, the World Health Organization (WHO) had listed As as a carcinogen and rereviewed the provisional guideline for drinking water, where concentration of As below 10 μg/l is considered to be safe (WHO 2011). Many technologies including adsorption, ion exchange, and reverse osmosis have been tested to remove As from water (Kartinen and Martin 1995). However, most of these techniques are expensive and non-eco-friendly. Use of microbial bioremediation has been one of the major scientific interests globally. Understanding of microbial metabolic system and their role in As cycling is very crucial for this purpose. This chapter focuses on the distribution of As and its fluxes across different ecosystems that are controlled by microbial metabolic pathways, which in turn depends on bioavailability and properties of organic matter in the environment that sustain these microbes.

5.2 Physicochemical Properties of Arsenic

Arsenic is present on Earth in the form of three prime allotropes, namely, black, gray, and yellow. The gray is the most abundant, while yellow is more volatile. Arsenic is monoisotopic, with only one isotope ^{75}As; however many As isotopes ranging from atomic number 60 to 92 have been synthesized (Shore et al. 2010). Due to its monoisotopic nature, As cannot be used by geochemists as a proxy for paleoenvironment-related studies or source-sink analyses. Compounds of As resemble that of phosphorus (P) compounds as both As and P belong to the same column of the periodic table. Arsenic mainly occurs in four states of oxidation: arsenate [As (V)], arsenite [As(III) +3], elemental [As(0)], and arsenide [As(III) -3]. In natural waters, like aquifers, ponds, geothermal spring, and rivers, As is predominantly found as arsenite [As(III)] or arsenate [As(V)] (Cullen and Reimer 1989). Among these, arsenite is more toxic to living forms than arsenate (Liu et al. 2001). Arsenite binds with thiol or vicinal sulfhydryl group receptor proteins and can affect cell signaling (Goyer and Clarkson 2001). It can also induce generation of reactive oxygen species (ROS) leading to cell damage and apoptosis (Liu et al. 2001).

Other than in sediments and mineral bearing lithospheric As, it is also present in atmosphere as small particle, produced from volcanic eruptions, industrial effluents, wind mobilization, and marine aerosol.

5.3 Forms of Arsenic in Sediments

Arsenic can be present in sediment in various inorganic and organic forms. The pentavalent inorganic forms are more abundant in lithosphere, including parts of various mineral forms (Kossoff and Hudson-Edwards 2012). Organic-rich arsenic forms are found during microbial leaching and uptake of inorganic As into their cell (Drewniak et al. 2010) which can be further transported in the food chain to higher trophic levels (Rahman et al. 2012).

The inorganic As is mostly present as minerals. Arsenic has more than 300 mineral forms, where oxides of arsenic and iron-bearing oxides of arsenic are mostly found. These minerals include arsenates, arsenites, arsenides, sulfides, sulfosalts, native elements, and metal alloys. Among these, sulfide-bound minerals (e.g., arsenopyrite, pyrite, realgar) and As(V) minerals (e.g., scorodite, beudantite, yukonite) are the most predominantly found sedimentary minerals (Kossoff and Hudson-Edwards 2012; Shrivastava et al. 2015) (Table 5.1). Other than these, arsenic is also found in toxic gaseous forms as arsine (AsH_3) (Kossoff and Hudson-Edwards 2012), released in highly reducing acidic environments like marshy lands (Kossoff and Hudson-Edwards 2012). Arsine is usually released with methylated organic forms of As.

Organic arsenic is mostly produced by living organisms, when they take up inorganic arsenic forms. Among these, aquatic organisms, e.g., microorganisms, plankton, and small and large fishes, as well as mammals play a crucial role. The inorganic As forms are methylated into less toxic organic forms monomethylarsine (MMA), dimethylarsine (DMA), and trimethylarsine (TMA) (Cullen and Reimer 1989). However, arsenobetaine (AsB) is reported to be the most commonly found natural organoarsenical which is absent in living organisms (Hopenhayn 2006). Other than these, organoarsine forms are reported from phospholipid extracts of many fishes (Lunde 1968) and non-phospholipids in humans (Amayo et al. 2011).

5.3.1 Distribution of Arsenic in Bengal Delta Plain (BDP)

The collision of Indian plate with that of Eurasian formed the Himalayas during Miocene (Umitsu 1993). The Bengal Fan was formed by deposition of sediments originating from rapid physical and chemical weathering of Himalayan rocks. This alluvial deposition further protruded as a clastic wedge into Bengal Basin (Alam 1989; Uddin and Lundbeg 1999). Lowering of sea level further along with erosion (Lindsay et al. 1991) and rapid deposition by the rivers such as Ganges, Brahmaputra, and Meghna resulted in the formation of BDP alluvial deposits (Acharyya et al. 2000). Although the first report of As contamination in groundwater

Table 5.1 Major mineral classes and forms of arsenic

Mineral class	Mineral name	Formula
Arsenite minerals	Arsenolite	As_2O_3
	Claudetite	As_2O_3
	Orpiment	As_2S_3
	Realgar	As_4S_4
Arsenate oxide	Arsenic pentoxide	As_2O_5
Fe-arsenate	Arsenopyrite	FeAsS
	Arseniosiderite	$Ca_2Fe_3O_2(AsO_4)_3{\cdot}3H_2O$
	Parasymplesite	$Fe_3(AsO_4)_2{\cdot}8H_2O$
	Pharmacosiderite	$K[Fe_4(OH)_4(AsO_4)_3]{\cdot}6.5H2O$
	Scorodite	$FeAsO_4{\cdot}2H_2O$
	Symplesite	$Fe_3(AsO4)_2{\cdot}8H_2$
	Yukonite	$Ca_7Fe_{12}(AsO_4)10(OH)_{20}{\cdot}15H_2O$
Fe sulfo-arsenates	Beudantite	$PbFe_3(AsO_4)(SO_4)(OH)_6$
	Tooeleite	$Fe_6(AsO_4)_4(SO_4)(OH)_4{\cdot}4H2O$
	Zýkaite	$Fe_4(AsO_4)_3(SO_4)(OH){\cdot}15H_2O$
Ca-Mg arsenates	Hörnesite	$Mg_3(AsO_4)_2{\cdot}8H_2O$
	Pharmacolite	$Ca(HAsO_4){\cdot}2H_2O$
	Adelite	$CaMgAsO_4OH$
Other metal arsenates	Annabergite	$Ni_3(AsO_4)_2{\cdot}8H_2O$
	Erythrite	$Co_3(AsO_4)_2{\cdot}8H_2O$
	Köttigite	$Zn_3(AsO_4)_2{\cdot}8H_2O$
	Mimetite	$Pb_5(AsO_4)_3Cl$

came from northern states of India in 1976, health-related concerns attracting scientific attention started only when reported from lower Gangetic plains of West Bengal, India (Garai et al. 1984). This was followed by the reports from Bangladesh having high As levels in groundwater. After those early reports on upper and lower Gangetic plains, levels of As had also been reported to be of concern from middle Gangetic plain (Chakraborti et al. 2003). The lower Gangetic deltaic plain, along with the deltaic plain of the rivers Brahmaputra and Meghna, forms highly fertile agricultural land known as Bengal Delta Plain (BDP). The high As in the entire course of river Ganga and in the BDP region comes from geogenic sources (Acharyya and Shah 2007). Initially it was postulated that the oxidation of As-rich pyrite causes release of As liberation in South Asia (Acharyya et al. 2000; Acharyya and Shah 2007); however subsequent reports show that chemical or microbiological reductive dissolution of As-bearing pyrite minerals causes such release (Bhattacharya et al. 1997; McArthur et al. 2001; Gadd 2004). The organic-rich peat beds and hydrocarbon resources buried in the sediments of BDP act as major organic source to sustain indigenous microbial communities (Bhattacharya et al. 1997; McArthur et al. 2001; Ghosh et al. 2014, 2015a; Whaley-Martin et al. 2016).

The mineralogy of the BDP deposits also plays a crucial role in As mobilization from sediments to aquifer water. The Fe(III) (e.g., pyrite, goethite, and ferrihydrite) or Mn(IV) (e.g., birmesite) (oxy)hydroxide-bound As present in these environment remains in sedimentary deposits under oxidized conditions (Smedley and Kinniburgh 2002). The framboidal pyrite deposits in the Gangetic plains and lack of sulfate in aquifer water are both signatures of diagenetic changes (Chowdhury et al. 2000).

Numerous studies had been undertaken to correlate sediment color and As flux in groundwater from aquifers of BDP (von Brömssen et al. 2007; Ghosh et al. 2015a). It has been clearly demonstrated that the aquifers with shallow gray sands, deposited during Pleistocene, are contaminated with As with levels above 10 μg/l. On the other hand the deep aquifers with brown-colored sediments deposited during late Pleistocene may be As safe (Biswas et al. 2012). These aquifers had also later been demonstrated to show contrasting features in terms of their organic matter deposits and microbiology. Thus the color perception had helped local drillers to avoid drilling boreholes in aquifers with possible As risk (Hossain et al. 2014). Although it is highly debatable, this can be a temporary measure to reduce high-risk health manifestations in local population.

5.4 Microbial Arsenic Cycling

Although As is highly toxic to all life forms, it can potentially act as an electron donor or acceptor in some microorganisms. Thus, microorganisms play a very important role in biogeochemical cycling and controlling the fluxes of As in diverse environments.

5.4.1 Microbial Arsenite Oxidation

5.4.1.1 Aerobic Arsenite Oxidation

The first aerobic bacterial As(III) oxidation was reported in 1918; however, this was overlooked till 1949 after which 15 new heterotrophic As(III) oxidizing bacteria were discovered (Green 1918; Turner 1949). The aerobic As(III) oxidizers can be divided into two main types: chemolithoautotrophic As(III) oxidizers (CAOs) and heterotrophic As(III) oxidizers (HAOs) (Oremland and Stolz 2003; Silver and Phung 2005; Stolz et al. 2010). During chemoautotrophic oxidation of As(III) to As(V), electron is sequestered into reduced oxygen to fix CO_2, whereas during heterotrophic As(III) oxidation, the As(V) is produced as less toxic form and ATP generates as a by-product (Fig. 5.2) (van den Hoven and Santini 2004). The key enzymes involved in both the types of As(III) oxidation is arsenite oxidase (Fig. 5.3). This enzyme was first isolated in the early 1990s (Anderson et al. 1992). The enzyme consists of two main subunits; larger one encoded by the gene *aioA* is a molybdopterin center with 3Fe/4S binding domain, and the smaller one encoded by the gene *aioB* is a Rieske protein (Lebrun et al. 2006; Lett et al. 2012). Other than these two, the operon encode three more proteins *aioR* (transcriptional regulator), *aioS* (sensor histidine kinase), and *aioX* (oxy-anion binding protein) (Lett et al. 2012). *Herminiimonas*

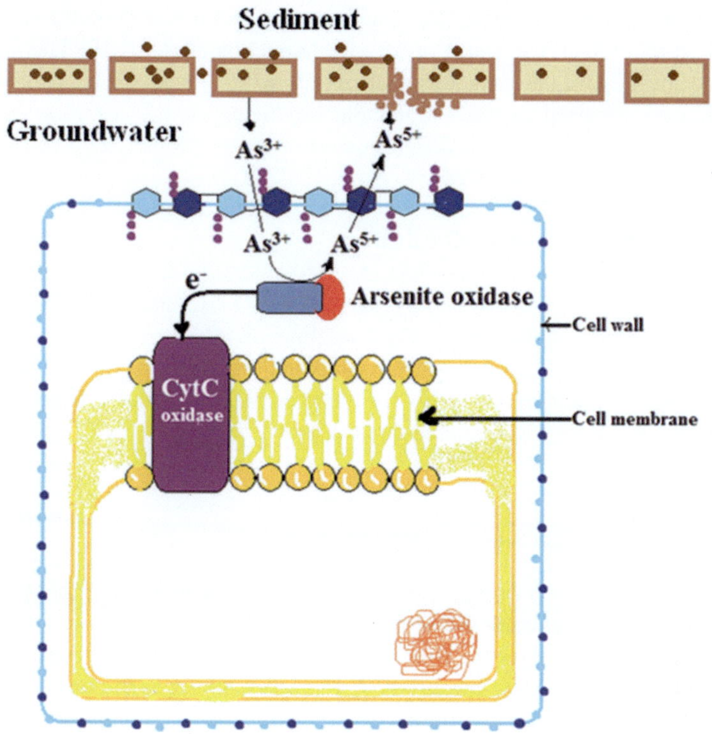

Fig. 5.2 Schematic representation of a bacterial cell oxidizing mobilized As(III) in groundwater to As(V) which get adsorbed into the sediment

Fig. 5.3 Model representation of the enzyme arsenite oxidase showing large and small subunit. (Modified from Silver and Phung 2005)

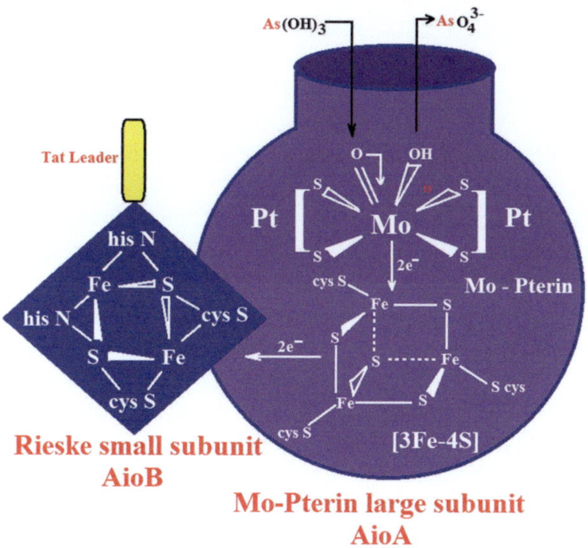

arsenicoxidans, an As(III)-oxidizing bacterium has been widely studied including the AioA protein which has a size of 96 kDa (Koechler et al. 2010). The *aioA* gene expressing this subunit has two consensus motifs which forms the basis of universal primers and widely used to study molecular phylogeny and distribution of As(III)-oxidizing bacterial groups across different environments, globally (e.g., Costello and Lidstrom 1999; Ghosh et al. 2014).

5.4.1.2 Anaerobic Arsenite Oxidation

In 2002, the first report of anaerobic As(III) oxidation was encountered in the bacterium *Alkalilimnicola ehrlichii* strain MLHE-1 isolated from Mono Lake, USA (Oremland and Stolz 2003). The coupling of nitrate reduction with As(III) oxidation is carried out by this bacterium in alkaline conditions of the lake through enzymes which are not encoded by the gene *aio*A, but genes closely related to arsenate reductase (Oremland and Stolz 2003).

5.4.2 Microbial Arsenate Reduction

The phosphate transporter proteins (Pit or Pst) on bacterial cell membrane act as an easy system for As(V) to enter the cell. Inside the cell, As(V) is reduced to As(III) in presence of glutathione or ferridoxine and excreted out via membrane efflux proteins ArsB or Acr3 (Rosen 1999). Moreover, As(V) can also act as the terminal electron acceptor in some bacterial groups (e.g., *Firmicutes*, *Gammaproteobacteria*, *Deltaproteobacteria*, and *Epsilonproteobacteria*) known as dissimilatory As(V)-reducing prokaryotes (DARPs; Oremland and Stolz 2003). This enzyme respiratory As(V) reductase (ARR; Fig. 5.4) has two subunits—the larger ArrA and the smaller ArrB (Saltikov and Newman 2003). The ArrA like AioA has a molybdenum-binding center and a 4Fe/4S–binding cluster. The smaller subunit ArrB can have three to four 4Fe/4S clusters.

5.4.3 Microbial Arsenic Methylation

Anaerobic methylation of As was first reported from the bacterium *Methanobacterium bryantii* strain MoH. Subsequently it was also reported from *Proteus* sp., *Escherichia coli*, *Flavobacterium* sp., *Corynebacterium* sp., and *Pseudomonas* sp. (Shariatpanahi et al. 1981). Along with *Flavobacterium* sp. and *Pseudomonas* sp., five other bacterial genera (*Achromobacter* sp., *Aeromonas* sp., *Alcaligenes* sp., *Enterobacter* sp., and *Nocardia* sp.) were also reported to convert As(V) to methylaronate and then to monomethyl arsenate (MMA) and dimethylarsenate (DMA). *Aeromonas* sp. and *Nocardia* sp. can further produce trimethyl arsenate (TMA) (Shariatpanahi et al. 1981). Such anaerobic methylation occurs in the presence of methylcobalamin (Ridley et al. 1977), catalyzed by the enzyme S-adenosylmethionine (SAM) methyltransferases and encoded by the *arsM* gene (Qin et al. 2006). This gene can be plasmid borne and thus limits its utility as a

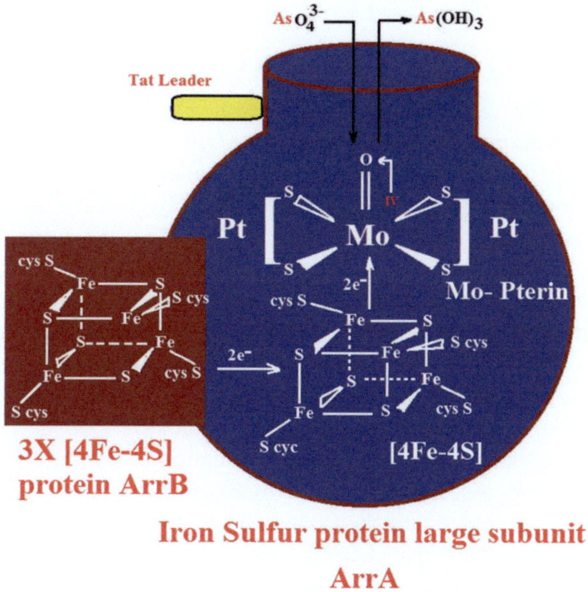

Fig. 5.4 Model representation of the enzyme arsenate reductase showing large and small subunit (Modified from Silver and Phung 2005)

molecular tool to study bacterial functional diversity. Methylation also leads to the formation of organoarsenicals {monomethylarsonic acid [MAs(V)], dimethylarsinic acid [DMAs(V) or cacodylate], and trimethylarsine oxide [TMAsO(V)]} which are much less toxic compared to inorganic forms (Challenger 1951) but have a crucial role in biogeochemical cycling of As in litho- and hydrospheres (Rensing and Rosen 2009).

5.4.4 Microbial Arsenic Demethylation

The chemical methylation and use of As in pesticide during the 1980s and 1990s had highly elevated the fluxes and resulted in the persistence of methylated As forms in the environment (Yoshinaga et al. 2011). However, prokaryotic demethylation of such large deposits has led to remediation of such pesticide-contaminated soil (Gao and Buran 1997; Maki et al. 2006), sludge (Sierra-Alvarez et al. 2006), sediment, seawater (Acharyya et al. 2000), lake water (Maki et al. 2006), and freshwater (Maki et al. 2009). Many bacterial strains such as *Mycobacterium neoaurum* (Lehr et al. 2003), *Pseudomonas putida* like strain KT2440 (Maki et al. 2006), *Burkholderia* sp. MR1, and *Streptomyces* sp. MR1 (Yoshinaga et al. 2011) had been reported to demethylate MMA into As(V).

5.5 Microbial Role in Mobilization of Arsenic in BDP Aquifers

It had been widely postulated since the late 1990s that in BDP, microbes are the key players for reductive dissolution of As-bearing Fe minerals under reducing conditions prevailing in these aquifers (Bhattacharya et al. 1997; Smedley and Kinniburgh 2002). Such desorption of As can take place by biotic and abiotic processes. A comparative study done by Lovely and Anderson (2000) had shown that some bacterial isolates (GS-15, *Escherichia coli*, and *Clostridium pasteurianum*) can reduce Fe(III) much faster and extensively in the presence of various organic compounds in comparison to reducing agents. Sedimentary organic matter plays a key role in sustaining microbial communities involved in the reductive dissolution of As-bound Fe (hydr)oxides under anoxic conditions of the BDP aquifers (Islam et al. 2004; Drewniak et al. 2010). It had been found that the predominance of Fe(III)-reducing bacteria *Geobacter* sp. in the lower Gangetic plains causes coupling As(V) reduction (Islam et al. 2004). The copulation of the two processes, oxidation of organic matter and reduction of As(V)-bearing Fe(III) oxides, was established based on further studies (e.g. Héry et al. 2010). Putative aerobic or denitrifying arsenate-reducing populations of *Pseudomonas*, *Elizabethkingia*, and *Pantoea* were reported along with the presence of iron-oxidizing *Sideroxydans* (Sultana et al. 2011).

In these oligotrophic aquifer waters, the production of different ligands and polysaccharide by indigenous microbial population for sequestration of organic matter additionally increases mineral dissolution. Bacterial siderophores are one of those known ligands for organic matter acquisition along with various heavy metals (Gadd 2004). Many studies had hypothesized As mobilization as a by-product of Fe (III) acquisition through siderophores from Fe (oxy)hydroxides (Mailloux et al. 2009). Siderophores from *Aspergillus niger*, *Mycobacterium* spp., *Pseudomonas azotoformans*, *Rhodococcus erythropolis*, *Albidiferax ferrireducens*, and *Rhizobium leguminosarum* had been reported to play key role in mineral dissolution and As mobilization (Lukasz et al. 2014). The presence of such bacterial groups has been later reported from many sites in BDP including aquifers (Sultana et al. 2011; Sarkar et al. 2013; Ghosh et al. 2014). Microbial metabolic processes can also lead to As dissolution as a by-product. The bacterium *Burkholderia fungorum* had been shown to produce acidic metabolite for weathering of minerals such as apatite thereby causing As release (Mailloux et al. 2009). Such mesophilic bacteria had been widely reported as dominant *Betaproteobacteria* in aquifer waters and sediments of BDP (Sultana et al. 2011; Sarkar et al. 2013; Ghosh et al. 2014).

5.5.1 Microbial Role in Immobilization of Arsenic in BDP Aquifers

The BDP aquifer waters are rich in Fe(II) oxy-hydroxides and remain in high concentration in anoxic systems (Chakraborty and Bhadury 2015). The tapped out water shows immediate oxidation and subsequent precipitation of Fe(III) oxides. Thus microbial oxidation plays a crucial role in controlling the As and Fe fluxes in

BDP aquifers. Microbial oxidation of As(III) to As(V) in BDP aquifers in different aquifer system had shown contrasting pictures in terms of predominance and diversity of arsenite (Ghosh et al. 2014). Phylogenetically diverse gene *aioA* (encoding larger subunit of arsenite oxidase; Fig. 5.4) has been detected in many bacterial genera isolated from BDP water and sediment samples including *Alcaligenes faecalis*, *Achromobacter* sp., *Pseudomonas* sp., *Thiobacillus ferrooxidans*, and *Thiobacillus acidophilus* (Inskeep et al. 2007). A high predominance of *Betaproteobacteria* due to mesophilic conditions has been mostly reported from molecular phylogenetic studies (Ghosh et al. 2014), with the predominance of genera such as *Hydrogenophaga*, *Acidovorax*, *Albidiferax*, *Bosea*, and *Polymorphum* (Ghosh et al. 2014). The phylogeny of functional gene *aioA* follows that of 16S rRNA inferring to their ancient origin (Ghosh et al. 2014). However conserved regions of the gene widely used to design primers have been reported to have inconsistency leading to a divergence from 16S rRNA taxonomic position of genera such as *Thermus* sp. and *Halorubrum* sp. (Sultana et al. 2012). This can be due to mutations in the primer binding region and thus can lead to phylogenetic diversity-based studies more biased (Sultana et al. 2012).

5.6 Geochemical Factors Affecting Arsenic Cycling in BDP

5.6.1 Inorganic Factors Affecting Arsenic Cycling in BDP

Although it is established that the aquifer waters of BDP are usually Ca–HCO_3 or Ca–Mg–HCO_3 types, Ca–Na–HCO_3-type and Na–Cl-type aquifers were also found locally. Interestingly, bicarbonates have overall dominant ion chemistry (Bhattacharya et al. 1997; Zheng et al. 2005; Biswas et al. 2012). The aquifers of BDP are largely classified into two different types based on their hydrogeochemical properties: the shallow Ca^{2+}–Mg^{2+}–HCO_3- type aquifers, with elevated levels of dissolved NH_4, Fe, and Mn and high P-extractable As in sediments, and, in contrast, the deep Na^+–HCO^{3-}type aquifers with low dissolved NH_4^+, Fe, and Mn and little phosphorus (P)-extractable As in sediments (Zheng et al. 2005). Such contrast had led scientists to conclude that P-extractable As present in sediments is the key source controlling mobilization of As into groundwater (Zheng et al. 2005). However, later it was observed that the local drillers practice a different way in order to target As-safe aquifers. The sand color of the aquifer came up as an indicator of As status. The gray sand aquifers (GSA) were found to be As contaminated (As >10 μg/l), and, in contrast, the brown sand aquifers (BSA) were As safe (As <10 μg/l; von Brömssen et al. 2007).

During temporal period of sea level regression, the GSAs were oxidized and thus had lower redox condition in comparison to BSAs (Umitsu 1993). This contrast leads BSAs not to reach a threshold of Fe (oxy)hydroxide reduction keeping them As safe (von Brömssen et al. 2007). However, there is some delineation of redox conditions in BSA aquifers due to Mn (oxy)hydroxides but not low enough for As mobilization (Biswas et al. 2012). On the other hand, the BSAs are rich in NH_4^+,

PO_4^{3-}, Fe, and As along with lower Eh indicating reductive dissolution of Fe (oxy) hydroxide (Biswas et al. 2012). This low reducing condition, coupled with microbial utilization of organic matter leading to dissolution of As-bearing Fe (oxy) hydroxides, increases flux of As in groundwater of GSAs. Thus, the organic matter composition plays a critical role in these aquifers for controlling microbial metabolic activities.

5.6.2 Organic Factors Affecting Arsenic Cycling in BDP

It had been widely hypothesized that in situ organic matter of BDP aquifers play key role in sustaining and shaping indigenous bacterial communities (Ravenscroft et al. 2001; McArthur et al. 2001; Rowland et al. 2009; Lawson et al. 2013; Whaley-Martin et al. 2016). The characteristics and bioavailability of organic compounds control indigenous microbial metabolism and subsequently rate of all geochemical processes. Despite this, there are very few studies available where the organic matter had been characterized and associated functions linked with.

5.6.2.1 Sedimentary Organic Factors Affecting Arsenic Cycling in BDP

The fluvial sediments of BDP are known to have low organic C content (TOC $< 1\%$; BGS/DPHE 2001; Ravenscroft et al. 2001; Zheng et al. 2005; Ghosh et al. 2015b). Most of the sedimentary OM were deposited during Pleistocene and Holocene depositions and can be mostly in recalcitrant form (Umitsu 1993; Ravenscroft et al. 2001; Dowling et al. 2002). These external OM can be pond derived, terrestrial, and anthropogenic which can percolate up to a depth of 100 m (Lawson et al. 2013). One of the most highlighted sources, which was previously not linked to As contamination in BDP aquifers, can be petroleum-derived natural hydrocarbons that seep into the shallow aquifers from deeper sediments of BDP (Rowland et al. 2009). The BDP region (both West Bengal and Bangladesh) is known to have reserves of unexplored gases and crude oils (Alam 1989; Ganguly 1997; Milici et al. 2002). The degradation of petroleum had, however, been reported in anoxic aquifers of other regions by indigenous anaerobic bacteria (Townsend et al. 2003) such as denitrifying bacteria (Ehrenreich et al. 2000). Such biodegradation of mature hydrocarbons had been reported from various shallow anoxic Fe(III)-reducing aquifers (Baedecker et al. 1993; Chapelle et al. 2002). These mature hydrocarbons were also reported from sediments of shallow aquifers of BDP (Rowland et al. 2009; Héry et al. 2010). Thus, the possibility of petroleum-derived mature hydrocarbons as electron donor for microbial-mediated Fe(III) reduction coupled with As mobilization in BDP aquifers needs further investigation.

Therefore, sedimentary organic carbon sources can be traced by bulk parameters of atomic C/N ratios, by stable C isotope ratios ($\delta^{13}C$ and $\delta^{15}N$), and by characterization of lipid biomarkers. However, isotopic signatures are not homogenous and are affected by digenetic and metabolic activities and assimilation of inorganic C from OM remineralization. Thus use of lipid biomarkers, which have longer shelf life, can be helpful in detecting the sources of OM in BDP aquifers.

5.6.2.2 Dissolved Organic Factors Affecting Arsenic Cycling in BDP

The sedimentary organic matter (OM) present in the soil zone seeps into the groundwater during aquifer recharging and thus contributes to the overall dissolved organic carbon (DOC) pool in aquifers. Intensive pumping out of water which reduces the water table generated a vacuum pull and increased dissolution of surface-derived OM in Cambodian and BDP aquifers (Lawson et al. 2013). The DOC-derived electron donors participate in redox reactions and transport of nutrients into the cell (Judd et al. 2006). Thus it affects subsurface biogeochemical processes and thereby sustains and shapes diverse microbial communities (Judd et al. 2006). Earlier studies on BDP aquifers suggest that DOC is mainly transported as a recharge from baseline of ponds, wetlands/marshlands, and/or subsurface clay plugs and lignite deposits in aquifer sediments (Lawson et al. 2013). The natural dissolved organic matter (DOM) in BDP aquifers acts as (i) a source of labile substrates for microbial humic acid and Fe(III) reduction, an As dissolution, and (ii) an electron acceptor (reduced humic acid) for microbial-mediated Fe (II) oxidation (Islam et al. 2004; Mladenov et al. 2010). Thus the DOM acts as an electron shuttle between different groups of microorganism and plays a very important role in controlling As fluxes in groundwater aquifers of BDP (Mladenov et al. 2010). It had been demonstrated that the dissolved concentration of As increases up to six times when incubated with goethite suspensions of sediments with pre-sorbed As and DOM of 25 mg/l compared to samples that do not have any DOM (Bauer and Blodau 2006). Thus, DOM has the potential to oxidize and reduce As in a short time (Bauer and Blodau 2006). Utilization of DOC of BDP groundwater by different bacterial groups had been studied (Sultana et al. 2012; Ghosh et al. 2014). However, there is paucity in terms of characterization of DOC and its bioavailability to microbial populations in groundwater aquifers.

5.7 Conclusion

Overall this chapter discusses how As contamination posed major human health risks globally. The high levels of geogenic As, as found in the sediments of BDP and got dissolved into aquifer waters, which is used for irrigation and as potable water sources, caused many health manifestations. Microbial cycling of As in those aquifers in the presence of typical organic matters had been proposed to be the main cause. Our knowledge of the combined effects of organic matter availability and microbial metabolism shaping the community structure in groundwater aquifers is not very clear to date from the context of As cycling. For example, the bioavailability and recalcitrance of particular classes of organic carbons and hydrocarbons can have contrasting effects on prokaryotic systems, and consequently, respective geogenic heavy metal-bearing minerals can exhibit various rates of dissolution and responses in a combination of different prokaryotic systems. Ultimately, the presence of such contaminants can affect the higher trophic levels of the food chain. Thus, in future studies, laboratory-based experiments involving bacterial strains isolated from different environments and grown in the presence of a combination of various organic matter sources can be in demand to fill the knowledge gap. Interdisciplinary studies involving modern techniques such as transcriptomics, metabolomics, and organic geochemical analysis using biomarkers and stable isotopes can significantly increase our understanding on the effect of organic carbon compounds on microbial As mobilization and immobilization. This ultimately may lead toward development of bioremediation technologies for As safe drinking water and water for agricultural purposes.

References

Acharyya SK, Shah BA (2007) Groundwater arsenic contamination affecting different geologic domains in India – review: influence of geological setting, fluvial geomorphology and quaternary stratigraphy. J Environ Sci Health A 42:1795–1805

Acharyya SK, Lahiri S, Raymahashay BC, Bhowmik A (2000) Arsenic toxicity of groundwater in parts of the Bengal basin in India and Bangladesh: the role of quaternary stratigraphy and Holocene sea-level fluctuation. Environ Geol 39:1127–1137

Alam M (1989) Geology and depositional history of Cenozoic sediments of the Bengal basin of Bangladesh. Palaeogeogr Palaeoclimatol Palaeoecol 69:125–139

Amayo KO, Petursdottir A et al (2011) Identification and quantification of arsenolipids using reversed-phase HPLC coupled simultaneously to high-resolution ICPMS and high resolution electrospray MS without species-specific standards. Anal Chem 83(9):3589–3595

Anderson GL, Williams J, Hille R (1992) The purification and characterization of arsenite oxidase from *Alcaligenes faecalis*, a molybdenum containing hydroxylase. J Biol Chem 267:23674–23682

Baedecker MJ, Cozzarelli IM, Eganhouse RP, Siegel DI, Bennett PC (1993) Crude oil in a shallow sand and gravel aquifer-III. Biogeochemical reactions and mass balance modelling in anoxic groundwater. Appl Geochem 8:569–586

Bauer M, Blodau C (2006) Mobilization of arsenic by dissolved organic matter from iron oxides, soils and sediments. Sci Total Environ 354:179–190

BGS, DPHE (2001) Arsenic Contamination of Groundwater in Bangladesh, in: Kinniburgh

Bhattacharya P, Chatterjee D, Jacks G (1997) Occurrence of arsenic contaminated groundwater in alluvial aquifers from Delta Plains, Eastern India: options for safe drinking water supply. Int J Wat Res Dev 13:79–82

Biswas A, Nath B, Bhattacharya P, Halder D, Kundu AK, Mandal U et al (2012) Hydrogeochemical contrast between brown and grey sand aquifers in shallow depth of Bengal Basin: consequences for sustainable drinking water supply. Sci Total Environ 431:402–412

Chakraborti D, Mukherjee S, Pati S, Sengupta MK, Rahman MM, Chowdhury UK et al (2003) Arsenic groundwater contamination in middle Ganga Plain, Bihar, India: a future danger. Environ Health Perspect 111:1194–1201

Chakraborty A, Bhadury P (2015) Effect of pollution on aquatic microbial diversity. In: Sukla LB, Pradhan N, Panda S, Mishra BK (eds) Environmental microbial biotechnology. Springer International Publishing, Cham, pp 53–75

Challenger F (1951) Biological methylation. Adv Enzymol Relat Sub Biochem 1:429–491

Chapelle FH, Bradley PM, Lovley DR, O'Neill K, Landmeyer JE (2002) Rapid evolution of redox processes in a petroleum hydrocarbon-contaminated aquifer. Groundwater 40:353–360

Chowdhury UK, Biswas BK, Chowdhury TR, Samantha G, Mandal B, Basu GC, Chanda CR, Lodh D, Saha KC, Mukherjee SK, Roy S, Kabir S, Quamruzzaman Q, Chakrabarti D (2000) Groundwater arsenic contamination in Bangladesh and West Bengal, India. Environ Health Prespect 8:393–397

Costello AM, Lidstrom ME (1999) Molecular characterization of functional and phylogenetic genes from natural populations of methanotrophs in lake sediments. Appl Environ Microbiol 65:50–66

Cullen WR, Reimer KJ (1989) Arsenic speciation in the environment. Chem Rev 89:713–764

Das HK, Mitra AK, Sengupta PK, Hossain A, Islam F, Rabbani GH (2004) Arsenic concentrations in rice, vegetables, and fish in Bangladesh: a preliminary study. Environ Int 30:383–387

Debnath M, Bhadury P (2016) Adaptive responses and arsenic transformation potential of diazotrophic Cyanobacteria isolated from rice fields of arsenic affected Bengal Delta Plain. J Appl Phycol 28:2777–2792

Dowling CB, Poreda RJ, Basu AR, Peters SL (2002) Geochemical study of arsenic release mechanisms in the Bengal Basin groundwater. Wat Res 38:1–18

Drewniak L, Matlakowska R, Rewerski B, Sklodowska A (2010) Arsenic release from gold mine rocks mediated by the activity of indigenous bacteria. Hydrometallurgy 104:437–442

Ehrenreich P, Behrends A, Harder J, Widdel F (2000) Anaerobic oxidation of alkanes by newly isolated denitrifying bacteria. Arch Microbiol 173:58–64

Emsley J (2001) Nature's building blocks: an A-Z guide to the elements, vol 43. Oxford University Press, Oxford, pp 513–529. isbn 0-19-850341-5

Gadd GM (2004) Microbial influence on metal mobility and application for bioremediation. Geoderma 122:109–119

Ganguly S (1997) Petroleum geology and exploration history of the Bengal Basin in India and Bangladesh. Indian J Geol 69:1–25

Gao S, Buran RG (1997) Environmental factors affecting rates of arsine evolution from mineralization of arsenicals in soil. J Environ Qual 26:753–763

Garai R, Chakraborti AK, Dey SB, Saha KC (1984) Chronic arsenic poisoning from tubewell water. J Ind Med Assoc 82:34–35

Ghosh D, Bhadury P, Routh J (2014) Diversity of arsenite oxidizing bacterial communities in arsenic-rich deltaic aquifers in West Bengal, India. Front Microbiol 5:602

Ghosh D, Routh J, Bhadury P (2015a) Characterization and microbial utilization of dissolved lipid organic fraction in arsenic impacted aquifers (India). J Hydrol 527:221–223

Ghosh D, Routh J, Därio M, Bhadury P (2015b) Elemental and biomarker characteristics in a Pleistocene aquifer vulnerable to arsenic contamination in the Bengal Delta Plain, India. Appl Geochem 61:87–98

Goyer RA, Clarkson TW (2001) Toxic effects of metals. Casarett & Doull's toxicology: the basic of poisons, 6th edn. McGraw Hill, New York

Green HH (1918) Isolation and description of a bacterium causing oxidation of arsenite to arsenate in cattle-dipping baths. Rept Direct Vet South Africa 6:593–599

Gulledge JH, O'Connor JT (1973) Removal of As(V) from water by adsorption on aluminum and ferric hydroxide. J Am Water Works Assoc 65:548

Héry M, van Dongen BE, Gill F, Mondal D, Vaughan DJ, Pancost RD, Polya DA, Lloyd JR (2010) Arsenic release and attenuation in low organic carbon aquifer sediments from West Bengal. Geobiology 8:166–168

Hopenhayn C (2006) Arsenic in drinking water: impact on human health. Elements 2:103–107

Hossain M, Bhattacharya P, Frape SK, Jacks G, Islam MM, Rahman MM et al (2014) Sediment color tool for targeting arsenic-safe aquifers for the installation of shallow drinking water tubewells. Sci Total Environ 493:615–625

Inskeep WP, Macur RE, Hamamura N, Warelow TP, Ward SA, Santini JM (2007) Detection, diversity and expression of aerobic bacterial arsenite oxidase genes. Environ Microbiol 9:934–943

Islam FS, Gault AG, Boothman C, Polya DA, Charnock JM, Chatterjee D, Lloyd JR (2004) Role of metal-reducing bacteria in arsenic release from Bengal delta sediments. Nature 430:68e71

Judd KE, Crump BC, Kling GW (2006) Variation in dissolved organic matter controls bacterial production and community composition. Ecology 87:2068–2079

Kabata-Pendias A, Pendias H (1984) Trace elements in soils and plants. CRC Press, Boca Raton. 315 pp

Kartinen EO, Martin CJ (1995) An overview of arsenic removal processes. Desalination 103:79–88

Koechler S, Cleiss-Arnold J, Proux C, Sismeiro O, Dillies MA, Goulhen-Chollet F (2010) Multiple controls affect arsenite oxidase gene expression in *Herminiimonas arsenicoxydans*. BMC Microbiol 10:53–65

Kossoff D, Hudson-Edwards KA (2012) The metabolism of arsenite. In: Santini JM, Ward SM (eds) Arsenic in the environment, vol 5. CRC Press, London, pp 1–23

Lawson M, Polya DA, Boyce AJ, Bryant C, Mondal D, Shantz A, Ballentine CJ (2013) Pond-derived organic carbon driving changes in arsenic hazard found in Asian groundwaters. Environ Sci Technol 47:7085–7094

Lebrun E, Santini JM, Brugna M, Ducluzeau AL, Ouchane S, Cothenet BS et al (2006) The Rieske protein: a case study on the pit falls of multiple sequence alignments and phylogenetic reconstruction. Mol Biol Evol 23:1180–1191

Lehr CR, Polishchuk E, Delisle MC, Franz C, Cullen WR (2003) Arsenic methylation by microorganisms isolated from sheepskin bedding materials. Hum Exp Toxicol 22:325–334

Lett MC, Muller D, Lievremont D, Silver S, Santini J (2012) Unified nomenclature for genes involved in prokaryotic aerobic arsenite oxidation. J Bacteriol 194:207–208

Lindsay JF, Holliday DW, Hulbert AG (1991) Sequence stratigraphy and the evolution of the Ganges–Brahmaputra Delta Complex. Am Assoc Petr Geol Bull 75:1233–1254

Liu SX, Athar M, Lippai I, Waldren C, Hei TK (2001) Induction of oxyradicals by arsenic: implication for mechanism of genotoxicity. Proc Natl Acad Sci U S A 98:1643–1648

Lovely DR, Anderson RT (2000) Influence of dissimilatory metal reduction on fate of organic and metal contaminants in the subsurface. Hydrogeol J 8(1):77–88

Lukasz D, Liwia R, Aleksandra M, Aleksandra S (2014) Dissolution of arsenic minerals mediated by dissimilatory arsenate reducing bacteria: estimation of the physiological potential for arsenic mobilization. Biomed Res Int 2014:841–892

Lunde G (1968) Analysis of arsenic in marine oils by neutron activation. Evidence of arseno organic compounds. J Am Oil Chem Soc 45(5):331–332

Mailloux BJ, Alexandrova E, Keimowitz AR, Wovkulich K, Freyer GA, Herron M, Stolz JF, Kenna TC, Pichler T, Polizzotto ML, Dong H, Bishop M, Knappett PSK (2009) Microbial mineral weathering for nutrient acquisition releases arsenic. Appl Environ Microbio 75:2558–2565

Maki T, Takeda N, Hasegawa H, Ueda K (2006) Isolation of monomethylarsonic acid-mineralizing bacteria from arsenic contaminated soils of Ohkunoshima Island. Appl Organomet Chem 20:538–544

Maki T, Hirota W, Ueda K, Hasegawa H, Azizur Rahman M (2009) Seasonal dynamics of biodegradation activities for dimethylarsinic acid (DMA) in Lake Kahokugata. Chemosphere 77:36–42

Mandal B, Suzuki K (2002) Arsenic round the world: a review. Talanta 58:201–235

McArthur JM, Ravenscroft P, Safiullah S, Thirlwall MF (2001) Arsenic in groundwater: testing pollution mechanisms for sedimentary aquifers in Bangladesh. Water Resour Res 37:109–117

Milici RC, Warwick PD, Attansai E, Wandrey CJ (2002) To sell or not sell: assessments of Bangladesh hydrocarbons. Oil and Gas Journal 100:24–28

Mladenov N, Zheng Y, Miller MP, Nemergut D, Legg T, Simone B, Hageman C, Rahman MM, Ahmed KM, McKnight DM (2010) Dissolved organic matter sources and consequences for iron and arsenic mobilization in Bangladesh aquifers. Environ Sci Technol 44:123–128

Oremland RS, Stolz JF (2003) The ecology of arsenic. Science 300:939–944

Qin J, Rosen BP, Zhang Y, Wang G, Franke S, Rensing C (2006) Arsenic detoxification and evolution of trimethylarsine gas by a microbial arsenite S-adenosylmethionine methyltransferase. Proc Natl Acad Sci U S A 103:2075–2080

Rahman MA, Hasegawa H, Lim RP (2012) Bioaccumulation, biotransformation and trophic transfer of arsenic in the aquatic food chain. Environ Res 35:116–118

Ravenscroft P, McArthur JM, Hoque BA (2001) Geochemical and palaeohydrological controls in pollution of groundwater by arsenic. In: Chappell WR, Abernathy CO, Calderon R (eds) Arsenic exposure and health effects IV. Elsevier Science Ltd, Oxford, pp 53–78

Rensing C, Rosen BP (2009) Heavy metals cycles (arsenic, mercury, selenium, others). In: Schaechter M (ed) Encyclopedia of microbiology. Elsevier, Oxford, pp 205–219

Ridley WP, Dizikes L, Cheh A, Wood JM (1977) Recent studies on biomethylation and biodemethylation of toxic elements. Environ Health Perspect 19:43–46

Rosen B (1999) Families of arsenic transporters. Trends Microbiol 7:207–212

Rowland HAL, Boothman C, Pancoast R, Gault AG, Polya DA, Lloyd JR (2009) The role of indigenous microbes in the biodegradation of naturally occurring petroleum, the reduction of iron, and mobilization of arsenite from West Bengal aquifer sediments. J Environ Qual 38:1598–1607

Roychowdhury T, Uchino T, Tokunaga H, Ando M (2002) Survey of arsenic in food composites from arsenic affected area of West Bengal, India. Food Chem Toxicol 40:1611–1621

Saltikov CW, Newman DK (2003) Genetic identification of a respiratory arsenate reductase. Proc Natl Acad Sci U S A 100:10983–10988

Sarkar A, Kazy SK, Sar P (2013) Characterization of arsenic resistant bacteria from arsenic rich groundwater of West Bengal, India. Ecotoxicology 22(2):363–376

Shah AQ, Kazi TG, Arain MB, Jamali MK, Afridi HI, Jalbani N, Baig JA, Kandhro GA (2009) Accumulation of arsenic in different fresh water fish species – potential contribution to high arsenic intakes. Food Chem 112:520–524

Shariatpanahi M, Anderson AC, Abdelghani AA, Englande AJ, Hugheres J, Wilkinson RF (1981) Biotransformation of the pesticide, sodium arsenate. J Environ Sc and Health 16:35–37

Shore A, Fritsch A, Heim M, Schuh A, Thoennessen M (2010) Discovery of the arsenic isotopes. At Data Nucl Data Tables 96:299–306

Shrivastava A, Ghosh D, Dash A, Bose S (2015) Arsenic contamination in soil and sediment in India: sources, effects, and remediation. Curr Pollut Rep 1:35–46

Sierra-Alvarez R, Yenal U, Field JA, Kopplin M, Gandolfi AJ, Garbarino JR (2006) Anaerobic biotransformation of organo-arsenical pesticides monomethylarsonic acid and dimethylarsinic acid. J Agric Food Chem 54:3959–3966

Silver S, Phung LT (2005) Genes and enzymes involved in bacterial oxidation and reduction of inorganic arsenic. Appl Environ Microbiol 71:599–608

Smedley PL, Kinniburgh DG (2002) A review of the source, behaviour and distribution of arsenic in natural waters. Appl Geochem 17:517–568

Stolz JF, Basu P, Oremland RS (2010) Microbial arsenic metabolism: new twists on an old poison. Microbe 5:53–59

Sultana M, Härtig C, Friedrich BP, Seifert J, Schlömann M (2011) Bacterial communities in Bangladesh aquifers differing in aqueous arsenic concentration. Geomicrobiol J 28:191–211

Sultana M, Vogler S, Zargar K, Schmidt AC, Saltikov C, Seifert J, Schlömann M (2012) New clusters of arsenite oxidase and unusual bacterial groups in enrichments from arsenic-contaminated soil. Arch Microbiol 194:623–635

Townsend GT, Prince RG, Suflita JM (2003) Anaerobic oxidation of crude oil hydrocarbons by the resident microorganisms of a contaminated anoxic aquifer. Eviron Sci Technol 27:5213–5216

Turner AW (1949) Bacterial oxidation of arsenic. Nature 164:76–77

Uddin A, Lundbeg N (1999) A paleo-Brahmaputra? Subsurface lithofacies analysis of Miocene deltaic sediments in the Himalayan-Bengal system, Bangladesh. Sediment Geol 123:239–254

Umitsu M (1993) Late quaternary sedimentary environments and landforms in the Ganges Delta. Sediment Geol 83:177–186

van den Hoven RN, Santini JM (2004) Arsenite oxidation by the heterotroph *Hydrogenophaga* sp. str. NT-14: the arsenite oxidase and its physiological electron acceptor. BBA-Bioenergetics 1656:148–155 von Brömssen M, Jakariya M, Bhattacharya P, Ahmed KM, Hasan MA, Sracek O et al (2007) Targeting low-arsenic aquifers in Matlab Upazila, southeastern Bangladesh. Sci Total Environ 379:121–132

Warren GP, Alloway BJ, Lepp NW, Singh B, Bochereau FJM, Penny C (2003) Field trials to assess the uptake of arsenic by vegetables from contaminated soils and soil remediation with iron oxides. Sci Total Environ 311:19–33

Whaley-Martin KJ, Mailloux BJ, van Geen A, Bostick BC, Silvern RF, Kim C, Ahmed KM, Choudhury I, Slater GF (2016) Stimulation of Microbially mediated arsenic release in Bangladesh aquifers by young carbon indicated by radiocarbon analysis of sedimentary bacterial lipids. Environ Sci Technol 50:7353–7363

WHO (2011) Guidelines for drinking-water quality, 4th edn. World Health Organization, Geneva, p 564

Yoshinaga M, Cai M, Rosen BP (2011) Demethylation of methylarsonic acid by a microbial community. Environ Microbiol 13:1205–1215

Zheng Y, van Geen A, Stute M, Dhar R, Mo Z, Cheng Z et al (2005) Geochemical and hydrogeological contrasts between shallow and deeper aquifers in two villages of Araihazar, Bangladesh: implications for deeper aquifers as drinking water sources. Geochim Cosmochim Acta 69:5203–5218

Bacterial Communities of Uranium-Contaminated Tailing Ponds and Their Interactions with Different Heavy Metals

6

Paltu Kumar Dhal

Abstract

Discharge of uranium (U) tailings and contaminant effluents from uranium ore extraction sites creates huge burdens of anthropogenic radioactivity and greatly alters the ecosystem. Remediating the environment from these contaminations thus becomes a huge responsibility of the industry. Microbe-based bioremediation has emerged as a potential alternative to hazardous mine waste management as well as removal of toxic contaminants efficiently from the environment. In order to formulate the bioremediation strategies effectively, it is essential to understand the inhabitant microbial community structure of mine sites and their metabolic role as related to those sites. In addition, deciphering microbial communities also helps us to understand the responsible biogeochemical cycling and food web dynamics of such sites. Advancement in different techniques that includes high-throughput DNA sequencing and different "omic" tools can provide details of microbial communities and their metabolic activity in contaminated environments. The present chapter will describe both culturable and unculturable microbial diversity, dynamics within the uranium tailing pond, and radionuclide-contaminated environment and their interaction with other heavy metals including uranium.

Keywords

Uranium tailings · Radionuclide · Microbial communities · Bioremediation

P. K. Dhal (✉)
Department of Life Science and Biotechnology, Jadavpur University, Kolkata, West Bengal, India
e-mail: pkdhal@lifesc.jdvu.ac.in

T. K. Adhya et al. (eds.), *Advances in Soil Microbiology: Recent Trends and Future Prospects*, Microorganisms for Sustainability 3,
https://doi.org/10.1007/978-981-10-6178-3_6

6.1 Introduction

Mining and purifying of uranium (U) for nuclear power and other activities have led to the environmental contamination in many nuclear energy-generating states (Dhal and Sar 2014). According to a recent report by Organization for Economic Co-operation and Development (OECD 2008), the global U production is likely to be increased to 40% by 2020, from 86,720 tonnes y^{-1} to 122,620 tonnes y^{-1}. Along with U, the presence of other radioactive elements (^{206}Pb, ^{230}Th, ^{226}Ra, etc.) and various other toxic heavy metals in U mine wastes poses a great environmental threat. Remediation of those contaminants can be only achieved by microbe-based bioremediation (Tabak et al. 2005; Wu et al. 2006). Extensive research has confirmed that microorganism-based cleanup strategy is one of the most appropriate and feasible alternatives to save our environment from any eventual disaster resulting from radionuclide contamination (Tabak et al. 2005). It is well established that microbe-based successful bioremediation strategies greatly depend on the understanding of the geochemical character of the contaminated sites along with the microbial communities involved in key physiological processes. As per the microbial communities are concerned, it showed demand on detail investigations to characterized its structure, metabolic potential, function, and interaction with inhabitant environments (Rittmann et al. 2006). Microorganisms in the terrestrial subsurface are responsible to mitigation of contaminants (Lloyd and Renshaw 2005; Gadd 2010; Dhal et al. 2011; Green et al. 2012). Owing to the complexity of these contaminated environments, increased attention has been focused to explore the indigenous microflora and their role in altering the mobility and toxicity of these metallic contaminants (Merroun and Selenska-Pobell 2008; Prakash et al. 2010; Pereira et al. 2012). In particular, considerable interests have been shown on understanding the composition of microbial community and their impact on the fate of U transport and availability, with the aim to exploit such microbial processes in bioremediation of large radionuclide- and metal-contaminated sites (Hwang et al. 2009; Lloyd and Gadd 2011). In recent years, extensive work has been undertaken to explore microbial communities with highly contaminated radioactive wastes and waste repositories including U mines and mine-affected environments using culture-independent as well as culture-dependent approaches. In order to decipher the importance of biogeochemistry of particular sites, it is essential to understand the phylogenetic diversity of the microbial ecology of those sites. This will help to get a clear idea of microbial community structure, community resilience, and their potential role in designing the systematic bioremediation strategies. The present chapter attempts to review indigenous bacterial diversity in U mine wastes and other radionuclide-contaminated sites. Attempts were also made to summarize the inhabitant microbial community within U and other heavy metal mine tailings around the world.

6.2 Critical Reviews and Analysis

Recently, considerable efforts have been made to decipher microbial diversity at various U mine and heavy metal-contaminated habitats using both culture-dependent and culture-independent approaches. Broad aim of most of these works was to elucidate microbial community composition and their interaction with U and other heavy metals to gain better insight in U biogeochemistry and the microbial role in bioremediation (Fredrickson et al. 2004; Fields et al. 2005; Brodie et al. 2006; Nedelkova et al. 2007; Michalsen et al. 2007; Akob et al. 2007; Barns et al. 2007; Hwang et al. 2009; Rastogi et al. 2010a, b; Islam et al. 2011; Mondani et al. 2011; Dhal et al. 2011; Dhal and Sar 2014; Kenarova et al. 2014; Leigh et al. 2014; Islam and Sar 2016). In general, it was observed that metal- and radionuclide-contaminated sites harbor large varieties of microorganisms organized in site-specific complex communities (Martinez et al. 2006; Merroun and Selenska-Pobell 2008; Rastogi et al. 2010b; Hemme et al. 2010; Dhal et al. 2011, Islam et al. 2011; Choudhary et al. 2012; Dhal and Sar 2014). In culture-dependent approaches, different bacterial strains isolated from these contaminated sites have also been characterized to decipher their interaction with metal/radionuclide (Selenska-Pobell et al. 1999; Panak et al. 2002; Suzuki and Banfield 2004; Nedelkova et al. 2007; Choudhary and Sar 2009, 2015; Choudhary et al. 2012; Islam and Sar 2016). It was observed that microorganisms present in U-contaminated sites can adapt themselves well to the local severe conditions and play important role(s) in affecting mobility of U and other heavy metals in the environment (Lovley et al. 1991; Merroun et al. 2005; Martinez et al. 2007; Choudhary and Sar 2011c; Choudhary et al. 2012; Choudhary and Sar 2015; Islam and Sar 2016).

6.2.1 Microbial Diversity in Uranium Mine Wastes and Other Radionuclide-Contaminated Sites

Microbial diversity studies in U and other radionuclide- and heavy metal-contaminated sites have been conducted for the last three decades or so. The most important driver of such studies is the fact that inhabitant microbes in highly contaminated sites play critical role in influencing the mobility and toxicity of such contaminants (Gadd 1992; Tabak et al. 2005; Merroun and Selenska-Pobell 2008). These studies were broadly focused into two distinct environments: (a) sites contaminated with U ore and mine wastes and (b) sites contaminated with radioactive waste generated from different nuclear activities.

Microbial diversity studies at U ore-/mine waste-contaminated sites have been conducted mainly in the USA, European countries, and Australia. Among the first report from the abandoned U mine at Rum Jungle, Australia, Goodman et al. (1981) indicated the presence of large and diverse microbial flora with *T. ferrooxidans* as the major bacteria consistently present with the sulfidic wastes. Microbial diversity in U mine waste heap near Ronneburg, Thuringia, Germany, was analyzed subsequently for the lithotrophic and chemoorganotrophic leach bacteria using the most probable number technique by Schippers et al. (1995). Selenska-Pobell et al. (2001) analyzed bacterial composition within two U waste piles from East Germany by rep-APD (repetitive primer-amplified polymorphic DNA), RISA (ribosomal intergenic spacer amplification), and 16S ARDREA (amplified ribosomal DNA restriction enzyme analysis) and showed predominance of *Acidithiobacillus ferrooxidans* and several *Pseudomonas* species. These authors hypothesized that different concentrations of heavy metals within U wastes are responsible for microbial diversity of the genus *Acidithiobacillus* in the contaminated environment.

16S rRNA gene-based microbial diversity study of samples from different depths of U mine waste pile at Haberland, Germany, or U depository site at Gunnison, Colorado, USA, revealed dominance of α-*Proteobacteria*, γ-*Proteobacteria*, and *Acidobacteria* (Selenska-Pobell et al. 2002; Selenska-Pobell 2002; Satchanska et al. 2004; Geissler and Selenska-Pobell 2005). Abundance of *Acinetobacter* spp. was observed in U waste from Steinsee Deponie B1, Germany (Radeva and Selenska-Pobell 2005). Reardon et al. (2004) showed predominance of *Proteobacteria* members *Oxalobacter*, *Duganella*, *Pseudomonas*, *Methylobacterium radiotolerans*, and *Alcaligenes* in surrogate minerals incubated in an acidic U-contaminated aquifer. Satchanska and Selenska-Pobell (2005) retrieved green non-sulfur bacteria and AD1 and OP11 divisions. Suzuki et al. (2005) observed the association of U(VI)-reducing *Geobacteraceae* and *Desulfovibrionaceae* members within sediment particles obtained from open pit of inactive U mine.

Rastogi et al. (2010b) characterized bacterial community of U-impacted soil samples using both high-density 16S rRNA microarray named PhyloChip and clone libraries and showed predominance of *Proteobacteria*, *Acidobacteria*, and *Bacteroidetes*. Several studies on 16S rRNA gene-based study in U-contaminated samples from Germany have found dominance of phylum *Acidobacteria* of total community (Selenska-Pobell 2002). The presence and abundance of phylum *Acidobacteria* sequences in U-contaminated subsurface sediments were noticed by Barns et al. (2007) too. The later investigators described the comprehensive and greatly expanded phylogeny of this phylum and noticed that sequences from U-contaminated sites were members of subgroups 1 to 3, 5, 6, 10, and 13. They have found that 1, 3, 4, and 6 are the most abundant subgroups in uncontaminated soils/sediments, while contaminated sediments contained a higher representation of new subgroups (13, 15, 18, etc.). Mondani et al. (2011) investigated the impact of U contamination on inhabitant bacterial community in soil content high in U. They have indicated that, in contrast to populations from nearby control samples, bacterial communities are composed of *Acidobacteria*, *Proteobacteria*, and other seven phyla-inhabiting uraniferous soils from the region of Bessines (Limousin, France) and exhibited specific fingerprints that were remarkably stable over time. Both the iron-reducing *Geobacter* and *Geothrix* and iron-oxidizing species such as *Gallionella* and *Sideroxydans* were detected in these regions.

Antunes et al. (2011) conducted a field study in a deactivated U mining area located in Cunha Baixa (Centre of Portugal) and evaluated the soil enzyme activities and potential nitrification across several contaminated sites. They also identified few parameters which can act as good indicators on soil microbial communities and soil functions. In India, microbial diversity within underground U ore deposits and host rocks from Jaduguda and Bagjata mines were first reported by Islam and Sar (2011a). It was indicated that U ore samples from Jaduguda were represented by lineages to uncultured and unclassified members of either *Pseudomonadales* of γ-*Proteobacteria* (35%) or *Chitinophagaceae* of *Bacteroidetes* (40%) or Gp4 of

Acidobacteria (50%), while, in contrast, major bacterial groups from Bagjata samples were mostly affiliated to cultivable genera of *Marinobacter* (32%), *Alcanivorax* (31%), *Agrobacterium* (85%), and *Acinetobacter* (70%). Latter authors (Islam et al. 2011) investigated microbial diversity of Banduhurang open cast U mine and indicated that there is a different in community structure in microbial diversity based on the site's contamination.

As reported, U ore samples showed diversity of bacteria of β-*Proteobacteria*, α-*Proteobacteria*, *Firmicutes*, *Bacteroidetes*, and *Actinobacteria* and uncultured *Acidobacteria*, *Chloroflexi*, and *Cyanobacteria*. Soil samples collected from mine periphery were dominated with uncultured *Acidobacteria* along with γ-*Proteobacteria*, β-*Proteobacteria*, α-*Proteobacteria*, δ-*Proteobacteria*, unclassified bacteria, uncultured *Bacteroidetes*, and others. Dhal and Sar (2014) investigated the microbial community in uranium-non-contaminated and uranium-contaminated samples in Jaduguda, Bagjata, and Turamdih uranium mines. They reported the dominance of *Acidobacteria*, *Bacteroidetes*, *Firmicutes*, and *Proteobacteria* followed by *Nitrospira*, *Deferribacteres*, and *Chloroflexi* in non-contaminated samples. In contrast, dominant bacterial group present in highly contaminated samples are *Proteobacteria* followed by *Acidobacteria* and *Bacteroidetes* and members of *Firmicutes*, *Chloroflexi*, *Planctomycete*, *Cyanobacteria*, *Actinobacteria*, and *Candidate* division.

Microbial community of uranium ore-bearing subsurface soil of Domiasiat in Meghalaya India was also investigated. Among the most dominant bacterial group α-*Proteobacteria* followed by *Acidobacteria* were observed in these samples (Kumar et al. 2013). In another report Chourey et al. (2013) investigated the active members of soil and groundwater microbial communities in the process of biostimulation at uranium- and nitrate-contaminated ORIFRC Area 2, USA, site using metaproteomic approaches. *Dechloromonas*, *Ralstonia*, *Rhodoferax*, *Polaromonas*, *Delftia*, *Chromobacterium* of β-*Proteobacteria*, and *Firmicutes* dominated as major groups at these sites.

Surface water-influenced soils with varying concentration of uranium and metal from former uranium mining district of Ronneburg, Germany, were targeted to evaluate the role of the uranium and other metals on microbial community in the long term, using 16S rRNA PhyloChip (Sitte et al. 2015). Their reports indicate that most abundant bacterial groups present in those sites are *Proteobacteria*, *Firmicutes*, *Actinobacteria*, *Bacteroidetes*, and *Acidobacteria*, and they vary greatly with the redox potential and secondary metal contaminants present in these sites. Radeva et al. (2013) investigated microbial community at abandoned U mining and milling sites of Bulgaria. They revealed *Proteobacteria*, *Acidobacteria*, and *Bacteroidetes* as the most abundant bacterial groups in these sites. Investigators (Maleke et al. 2014) also tried to evaluate if there is any change in the indigenous bacterial community during the soluble uranium bioremediation using biostimulation system. Later in a work on the ORIFRC site, the flexibility of groundwater functional microbial communities was reported during injection of emulsified vegetable oil (EVO) in uranium-bearing contaminant plume (Zhang et al. 2015). Authors also tried to identify the key genes and bacterial groups involved in EVO biodegradation in this environment. Brzoska and Bollmann (2016) investigated the dynamics of artificially prepared microbial consortia because of the acidic pH and elevated uranium. Microbial consortium was composed artificially with *Caulobacter* sp. OR37, *Asinibacterium* sp. OR53, *Ralstonia* sp. OR214, and *Rhodanobacter* sp. OR444. All those bacteria were isolated from a uranium-contaminated and acidic subsurface sediment of IFRC in Oak Ridge (TN, USA).

Microbial diversity within radioactive contaminated sites has been elucidated mainly under the bioremediation research program of US Department of Energy (US-DOE). Under this program, extensive work has been carried out to decipher subsurface microbial communities within extremely contaminated (by nitric acid, U, Tc, Ni, Hg, Cr, Al, etc.) Oak Ridge Field Research Center (ORFRC) and high-level nuclear waste-contaminated sediments from Hanford Site, Washington. In order to formulate an effective strategy for bioremediation of radionuclide contamination, tremendous research has been conducted to understand the composition of microbial communities and its metabolic potential in highly contaminated environments using mainly cultivation-independent methods (Akob et al. 2007, 2011; Lovley 2003; Marsili et al. 2007; Beyenal et al. 2004; Cardenas et al. 2008). Distribution, diversity, and physiology of microbial communities within U-contaminated US-DOE sites have been reported by Petrie et al. (2003). Predominance of iron (III)-reducing bacterial community, particularly the *Geobacteraceae* and *Anaeromyxobacter* in pristine and contaminated sediments, respectively, has been observed. Fredrickson et al. (2004) reported that radionuclide-contaminated vadose sediment was represented by lower number of viable heterotrophic bacteria mostly affiliated to *Arthrobacter*, high G+C Gram-positive bacteria, *Rhodococcus* and *Nocardia*, and highly radiation-resistant *Deinococcus radiodurans*. Fields et al. (2005) observed high bacterial diversity with *Arthrobacter* and *Novosphingobium* as frequently detected bacterial groups in Field Research Center (FRC) located within the Y-12 Security Complex near Oak Ridge, TN. Contaminated sites were dominated by *Azoarcus* and *Pseudomonas*. Akob et al. (2007) showed high abundance of *Proteobacteria* (represented by the genera *Sphingomonas*, *Acidovorax*, *Acinetobacter*, *Alcaligenes*, and *Ralstonia*) in total and metabolically active fractions of the microbial community in U-contaminated ORFRC sediment.

Functional gene array method was adopted to analyze microbial communities of U-contaminated groundwater and to detect responsible genes for metal resistance, organic contaminant degradation, carbon and nitrogen cycle, and sulfate reduction (Waldron et al. (2009)). Nostrand et al. (2011) investigated microbial communities from three monitoring wells of highly U-contaminated aquifer at Oak Ridge, TN, with a high-density comprehensive functional gene array named GeoChip. They revealed that functional populations of Fe (III)-, nitrate-, and sulfate-reducing bacteria dominated in the active U(VI) reduction phase. Investigation by Green et al. (2012) revealed the long-term effect of U and nitrate on microbial community of terrestrial subsurface at the ORIFRC site in Oak Ridge, TN, and reported the presence of *Rhodanobacter* as the dominant bacterial member. Cho et al. (2012) have tried to understand the link between bacterial diversity and geochemistry in U-contaminated groundwater from ORFRC site by assessing microbial communities by 16S rDNA gene-based clone library analysis. High concentrations of contaminants present in well FW113-47 stimulated the growth of organisms capable of reducing U (*Shewanella* and *Pseudomonas*), nitrate (*Pseudomonas*, *Rhodanobacter*, and *Xanthomonas*), and iron (*Stenotrophomonas*). Mosher et al. (2012) investigated bacterial community structure during bioremediation using lactate amended in a continuous-flow reactors containing Cr (VI)-contaminated groundwater of Hanford area. Microbial community composition of initial groundwater sample dominated with the sequences were classified as members of phyla *Clostridia* > unclassified β-*Proteobacteria* > unclassified γ-*Proteobacteria*.

Details of microbial diversity were studied at the Chernobyl Nuclear Power Plant after the meltdown and subsequent explosion of one of the reactors at that site (Ragon et al. 2011). Later Chapon et al. (2012) reported the presence of bacterial communities of phylum *Firmicutes*; *Actinobacteria*; α-, β-, and γ-*Proteobacteria*; and *Bacteroidetes* in radioactive contaminated soils using both of culture-dependent and culture-independent approaches. Results showed that both contaminated and relatively less contaminated soils showed a wide diversity of bacteria, thus indicating the exposure to radionuclides for long term not leads to the loss of the diversity of bacterial community. Another group from the USA (Konopka et al. 2013) reported the microbial activity of the aquifer in 300 area of the Hanford Site. Their reports indicated the presence of bacterial group dominated with members from the *Proteobacteria* (particularly the alpha, beta, and gamma clades) followed by *Bacteroidetes* and *Actinobacteria*.

Diversity of *Archaebacteria* was investigated in several U-contaminated sites. Among the earlier studies, Fuchs et al. (1995) isolated thermoacidophilic metal-mobilizing *Archaea*, *Metallosphaera prunae*, from U mine in Thuringia, Germany. *Methanobacteria* of *Euryarchaeota* was predominant at those sites (Suzuki et al. 2005). *Methanobacterium subterraneum*, a methane-producing *Archaea*, was isolated from groundwater of Simpevarp nuclear power plant, Southeast Sweden (Kotelnikova et al. 1998). Archaeal diversity of different metal-contaminated site has also been investigated (Takai et al. 2001; Stein et al. 2002). Reduced capability of hyperthermophilic crenarchaeon *Pyrobaculum islandicum* on transformation of U (VI) to U (IV) was investigated (Kashefi and Lovely 2000). Geissler (2007) investigated the effect of uranyl or sodium nitrate on archaeal diversity in U mining waste of Johanngeorgenstadt, Germany, under aerobic and anaerobic conditions. The result indicated the abundance of mesophilic *Crenarchaeota* (64%) and the crenarchaeal (36%) group 1.1b in the untreated samples, while uranyl-treated samples were dominated with uncultured *Crenarchaeota* of the mesophilic group 1.1b. Porat et al. (2010) have investigated archaeal communities from mercury- and U-contaminated freshwater stream sediments located in the vicinity of Oak Ridge, TN, USA. These investigators indicated that *Crenarchaeota* comprised 76% of total community while the remaining 24% were from *Euryarchaeota*. Diversity of archaeal communities inhabiting the abandoned mining and milling complex "Buhovo" and the "Sliven" mine environments impacted was also investigated (Radeva et al. 2014), and data revealed that the only group that dominated this site is 1.1b/*Nitrososphaera*.

6.2.2 Microbial Diversity in Uranium Mine Tailings Sites

Miller et al. (1987) from US Geological Survey and Colorado School of Mines studied the ecological aspects of microorganisms inhabiting U mill tailings using conventional culture-based techniques. These investigators had observed the presence of *Arthrobacter* and *Bacillus* as major bacterial genera including sulfate-reducing bacteria and some fungi within the tailing environment. In another report Silver (1987) reported the presence of the Fe-oxidizing bacteria in the Nordic U tailing deposit, Canada. This investigator has reported the presence of iron-oxidizing bacteria within the top 2 m and near the water table-capillary fringe of the vegetated Nordic U deposit. Several studies have been carried out considering the role of microorganism on influencing the behavior of radionuclides and metals in tailings and other mine wastes (Ivanova et al. 2000; Chang et al. 2001; Selenska-Pobell et al. 2001; Elias et al. 2003; Landa 2005; Radeva and Selenska-Pobell 2005; Wolfaardt et al. 2008). Abundance of *Acinetobacter* spp. was observed in U waste from Steinsee Deponie B1, Germany (Radeva and Selenska-Pobell 2005). Elias et al. (2003) investigated the different responsible microbiological and geochemical factors for in situ U redox reactions in Shiprock, New Mexico, USA, U mill tailings and observed higher abundance of sulfate- and nitrate-reducing microorganisms. They have also observed preferred reduction of nitrate which slowed down the reduction rate of sulfate, Fe (III), and U. In another study, Ivanova et al. (2000) analyzed 16S rRNA and *amo*A genes and observed predominance of ammonia-oxidizing bacteria (*Nitrosomonas* sp.) in the groundwater near the same site. Predominance of *Nitrosomonas* was also noticed by Radeva and Selenska-Pobell (2005) in groundwater of Shiprock mill tailings as well. Both these studies indicated predominance of nitrogen-metabolizing bacteria in the Shiprock tailings site, and based on these findings, Elias et al. (2003) suggested that nitrogenous compounds should be removed from such contaminated sites prior to the stimulation of U-reducing microbial community. Analysis of dissimilatory sulfite reductase (DSR) genes and phospholipid fatty acid (PLFA) profile of sulfate-reducing bacteria (SRB) from U Mill Tailings Remedial Action (UMTRA) disposal cell in Shiprock revealed predominance of *Desulfotomaculum* and *Desulfotomaculum* and *Nitrospira* division (Chang et al. 2001). The authors further suggested these organisms's possible involvement in the natural attenuation of U. Radeva and Selenska-Pobell (2005), however, demonstrated lower abundance of SRB in similar type of sample from Shiprock tailings site. The latter investigators compared bacterial communities of three different tailings sites and observed that *Nitrospina* were dominated in U mill tailings of Schlema/Alberoda, Germany, while *Pseudomonas* and *Frateuria* spp. is abundant in U mill tailings of Shiprock. In other U mill tailings at Gittersee/Coschütz, the dominant bacterial groups are *Proteobacteria* and *Cytophaga-Flavobacterium-Bacteroides*. Wolfaardt et al. (2008) characterized the distribution and diversity of microbial communities in mine tailings site of Rabbit Lake U mine, Canada, and observed a difference in community composition (measured by metabolic profile) with depth. These investigators have also observed the ability of inhabitant bacteria to develop biofilm and sustain their activity at broad range of pH on tailings water and suggested that resilience and adaptive nature of these microbial communities could have significant potential in long-term geochemical evolution of the tailings management facilities (TMF). Bondici et al. (2013) investigated the microbial diversity of tailings core samples of uranium mine tailings from Key Lake, Saskatchewan, Canada. This tailing pond is characterized as low permeability, high pH tailing pond. Report indicates those dominant microbial groups present in those samples are *Gemmatimonadetes*, *Chloroflexi*, *Deferribacteres*, *Deinococcus-Thermus*, *Verrucomicrobia*, *Aquificae*, and *Thermotogae*. In another report from the same group, Khan et al. 2013 investigated the microbial diversity from the same sites. Unlike the previous investigation, the samples collected in this study were mainly from uranium mine-water tailings interface, and as indicated *Firmicutes* was the most abundant bacterium present in these sites. In India, microbial diversity of Jaduguda U mine tailing samples was investigated by Dhal and Sar (2014). They revealed that this site showed predominance of phyla *Proteobacteria* and *Acidobacteria* followed by *Actinobacteria*, *Chloroflexi*, *Cyanobacteria*, *Firmicutes*, and *Planctomycete* and genera *Incertae sedis* OP10. The presence of *Crenarchaeota* in tailing samples was also reported. Yan and Luo (2015) reported the microbial diversity of U mill tailings from Southeastern China where *Serratia* sp. of *Proteobacteria* was the dominant strain in those sites. Recently Yan et al. (2016) reported the structural and functional diversity of microbial community in different depths from uranium-contaminated and uranium-non-contaminated areas of Southern China using metagenomic approaches where *Proteobacteria* and *Actinobacteria* were dominant in uncontaminated soils and *Robiginitalea*, *Microlunatus*, and *Alicyclobacillus* were abundant in radioactive contaminated soil. Sànchez-Castro et al. (2017) isolated two uranium-resistant strains (*Arthrobacter* sp. and *Microbacterium oxydans*) from uranium mill tailing repository sites of Bessines-sur-Gartempe (Limousin, France) and evaluated their potential role for U bioremediation

6.2.3 Microbial Interaction with Metals

Microbial interaction with metals and radionuclides has inspired environmental microbiologist for a long time. Compared to other living organisms, microbes have coexisted with metals since the early history which possibly resulted in the recruitment of wide range of divalent or transition metals in all aspects of microbial growth, metabolism, and differentiation (Gadd 2010). The bacterial system interact with metals through various mechanisms: (i) biosorption/intracellular accumulation, (ii) complexation by metal-binding molecules, (iii) enzymatic precipitation and biomineralization, and (iv) oxidation and reduction reactions (Gadd 2004). Emergence of such interactive mechanisms not only facilitates environmental fate of metals by their redox transformation or altered mobility/solubility and toxicity but also ensures the adaptation of microorganisms in a changing environment. All such interaction mechanisms that eventually alter the toxicity and mobility of metals and other radionuclides open a new area of research to their applications in bioremediation of metal and radionuclide contaminants (Barkay and Schaefer 2001).

Biosorption and bioaccumulation of metals by bacteria are the most frequently used mechanisms responsible for soluble metal concentration in the environment. Biosorption is a metabolism-independent metal and radionuclide's accumulation capability by microbial cells. On the contrary, bioaccumulation is metabolism-dependent intracellular metal accumulation potential of live cells that occurs in two stages, first the rapid metal binding on the cell surface followed by intracellular accumulation at a slower rate by active cell membrane transporter (Choudhary and Sar 2011a). Metal accumulation by biosorption or bioaccumulation is the first step of metal nucleation, precipitation, and biomineral formation. One of the early groups that worked on microbial metal accumulation was at Oak Ridge National Laboratory, USA. The group led by G W Strandberg has demonstrated that accumulation of metals in microbial cells occurs by both metabolic and non-metabolic processes, leading to either extracellular or intracellular precipitation (Strandberg et al. 1981). The work on biosorption continued to expand in the early 1980s with the development of freely suspended and immobilized biomass types as biosorbents for the treatment of contaminated wastewaters (Tsezos and Volesky 1981). Tsezos et al. (1997) demonstrated metal biosorption by *Arthrobacter* spp., *Alcaligenes eutrophus*, and *P. mendocina*. The role of bacterial surface structures on cell wall carbohydrate polymers or proteinaceous surface layer in metal binding was observed by T J Beveridge (Douglas and Beveridge 1998). Binding capability of U by various inhabitant-isolated bacteria from U-contaminated sites were investigated (Andres et al. 2001; Panak et al. 2002; Francis et al. 2004; Nakajima and Tsuruta 2004; Merroun and Selenska-Pobell 2008; Kazy et al. 2009; Choudhary et al. 2012). Results of all these studies demonstrated a species-specific U biosorption property which is affected by the chemical nature and pH of the solution, cell's physiology, as well as the presence of soluble polymer. Phosphoryl residues of phospholipids and lipopolysaccharide (LPS) in the outer membrane of *E. coli* K-12 and *Pseudomonas aeruginosa* have been observed to be the most probable metal-binding sites (Langley and Beveridge 1999; Choudhary et al. 2012). A study on the U adsorption on the cell surface of *Bacillus subtilis* under acidic conditions demonstrated that carboxylic and phosphate groups act as binding ligands and are responsible for coordination of the radionuclide (Fowle et al. 2000; Kelly et al. 2002). Gorman-Lewis et al. (2005) have suggested that in acidic condition adsorption onto bacterial cell surfaces because of electrostatic forces as well as covalent binding with their cell surface. Merroun and Selenska-Pobell (2008) revealed role of carboxyl and phosphate groups for binding the U by *Bacillus sphaericus* JG-A12 isolated from a U-contaminated environment. Various spectroscopic and microscopic techniques such as time-resolved laser-induced fluorescence spectroscopy (TRLFS), energy-dispersive X-ray (EDX) analysis, transmission electron microscopy (TEM), scanning electron microscopy (SEM), X-ray diffraction (XRD), Fourier transform infrared (FTIR) spectroscopy, and X-ray photoelectron spectroscopy (XPS) analysis have been used in these studies (Chojnacka 2010; Merroun and Selenska-Pobell 2008; Choudhary et al. 2012). TEM observations have revealed that U absorbed by Gram-positive bacteria

Mycobacterium smegmatis is deposited in cytoplasm or cell wall (Andres et al. 1994), whereas, in Gram-negative *P. fluorescens*, deposited U formed platy crystal structures within periplasmic and cell envelope region (Krueger et al. 1993). Formation of needle-shaped U deposits in cell envelope regions has been observed in several studies (Choudhary et al. 2012). In recent years, extended X-ray absorption fine structure (EXAFS) measurements have been applied to decipher the chemical mechanism of U sequestration by a number of bacteria (Merroun and Selenska-Pobell 2008; Templeton and Knowles 2009).

Microorganisms can mobilize/immobilize metals by reduction and oxidation processes (Lovley et al. 1991). Reduction is one of the most important chemical transformations catalyzed by microorganisms, which affect the solubility of metallic elements. Following reduction, solubility of many of these metal ion increases as they are transformed to lower oxidation state. In contrast to this, the solubility of other metals and radionuclides decreases as they are reduced to lower oxidation state, resulting in their immobilization (Lovley and Coates 1997). These interactions facilitate alteration in mobility and toxicity of metals by redox-mediated biotransformation, intracellular or extracellular complexation or sequestration, induction of metal precipitation, and biomineralization or change in metal speciation caused by microbially induced redox alteration (Lovley 1993; Bosecker 1997; Selenska-Pobell et al. 1999; McLean and Beveridge 2001; Lack et al. 2002; Merroun et al. 2002; Suzuki and Banfield 2004; Beller 2005; Merroun et al. 2005; Pollmann et al. 2005; Jroundi et al. 2007; Nedelkova et al. 2007; Merroun and Selenska-Pobell 2008; Sivaswamy et al. 2011; Choudhary and Sar 2011b; Yi and Lian 2012). The group led by Selenska-Pobell investigated the interaction with metals including U by bacterial isolates from U-rich site. Identification of acidophilic Fe- and S-oxidizing bacteria *A. ferrooxidans*, *A. thiooxidans*, *A. acidophilus*, and *Leptospirillum ferrooxidans* from U deposits and mineral heaps has been reported (Cerdá et al. 1993; Panak et al. 1998). Panak et al. (1998) investigated interaction of *A. ferrooxidans* and *Desulfovibrio desulfuricans* with U. It was observed that U mine isolate *A. ferrooxidans* ATCC 33020 can accumulate intracellular U. Merroun and Selenska-Pobell (2001) and Hafez et al. (2002) reported U sorption by a number of *Bacillus* and *Acidithiobacillus* strains isolated from radionuclide- and U-contaminated waste. A number of studies have demonstrated the U cell wall interactions of Gram-positive bacteria (i.e., *Bacillus*) and Gram-negative bacteria (*Shewanella putrefaciens*, *Sphingomonas* sp. S15-S1) (Fowle et al. 2000; Kelly et al. 2002; Gorman-Lewis et al. 2005; Merroun and Selenska-Pobell 2008; Yi and Lian 2012), and biosorption ability of the bacteria varied with their Gram characteristics.

With respect to characterization of U and other heavy metal accumulation by bacteria isolated from U mine/U-contaminated sites, several studies were conducted to isolate bacterial strains as their pure culture (Selenska-Pobell et al. 1999; Suzuki and Banfield 2004; Martinez et al. 2006; Beazley et al. 2007; Nedelkova et al. 2007; Tsuruta 2007; Choudhary and Sar 2009; Islam and Sar 2011; Yi and Lian 2012). Bacterial cells are found to have U bioaccumulation potential mainly by metabolic-independent process. It was reported that U transported into microbial cells because of the increased membrane permeability (Suzuki and Banfield 1999, 2004).

P. aeruginosa, *B. subtilis*, and *Chryseomonas* MGF48 have shown to immobilize U by cellular uptake (Strandberg et al. 1981; Fowle et al. 2000; Malekzadeh et al. 2002; Choudhary and Sar 2011b). Merroun et al. (2003, 2005) have reported accumulation of U immobilized by chelating with intracellular polyphosphates. Immobilization of U can also occur by precipitation with metabolites released by cells (Sivaswamy et al. 2011). *Citrobacter* sp., *A. ferrooxidans*, *B. sphaericus*, and *Acinetobacter johnsonii* have been shown to remove U from water by releasing phosphate/polyphosphate to the water (Yong and Macaskie 1998; Boswell et al. 1999; Merroun et al. 2002; Knopp et al. 2003). Using this mechanism, removal of U (VI) was observed through formation of U phosphate minerals in aerobic, anaerobic environments (Beazley et al. 2007, 2009; Martinez et al. 2007). Nilgiriwala et al. (2008) cloned *pho*K gene from *Sphingomonas* sp. for enhance production of phosphatase gene and demonstrated the potential of U bioprecipitation.

Bioreduction of soluble U (VI) to insoluble U (IV) was proposed to mitigate the spread of U under anoxic condition (Lovley et al. 1991; Gorby and Lovley 1992). Dr. Lovley and his team has done extensive work on Fe (III)-reducing bacteria for enzymatic U reduction. As reviewed by Merroun and Selenska-Pobell (2008), more than 25 phylogenetically diverse prokaryotes showed tremendous potential in the reduction of this radionuclide and include sulfate-reducing bacteria, Fe(III)-reducing bacteria, hyperthermophilic archaea, thermophilic bacteria, fermentative bacteria from *Clostridium* spp., acid-tolerant bacteria, radioresistant bacteria (*Deinococcus radiodurans* R1), as well as myxobacteria (*Anaeromyxobacter* spp.) (Lovley et al. 1991; Lovley and Phillips 1992; Kieft et al. 1999; Kashefi and Lovely 2000; Coates et al. 2001; Shelobolina et al. 2004; Suzuki et al. 2005; Suzuki and Suko 2006; Wu et al. 2006; Francis et al. 2008). Investigators (Fletcher et al. 2010) also demonstrated the reduction of U (VI) by Gram-positive *Desulfitobacterium* spp. These investigations have shown that U (VI) reduction can be enzymatic events and humic acid-mediated electron transfer as well.

6.3 Conclusion

In contrast to other developed and nuclear power countries, knowledge on microbial diversity and their role in U mine sites from India is quite fragmentary, although India is targeting higher power generation through U-based reaction that essentially required enhanced U ore process. In order to develop sustainable mine waste remediation strategies, understanding of the geomicrobiology of such sites is imperative. Along with that, knowledge on microbial potential and their role in dealing with such toxic elements (U and other metals) is not complete, particularly, in consideration with the vast metabolic and genetic diversity of microbial world and their nature of non-cultivability. Therefore, there is a strong demand for further investigation of indigenous microflora of U mine tailings and other mine sites as well as on isolating pure culture bacteria from highly contaminated habitats, deciphering their physiology and metabolic capabilities, and understanding their potential in bioremediation.

References

Akob DM, Mills HJ, Kostka JE (2007) Metabolically active microbial communities in uranium-contaminated subsurface sediments. FEMS Microbiol Ecol 59:95–107

Akob DM, Kerkhof L, Küsel K, Watson DB, Palumbo AV, Kostka JE (2011) Linking specific heterotrophic bacterial populations to bioreduction of uranium and nitrate in contaminated subsurface sediments by using stable isotope probing. Appl Environ Microbiol 77(22):8197–8200

Andres Y, MacCordick HJ, Hubert JC (1994) Binding sites of sorbed uranyl ion in the cell wall of *Mycobacterium smegmatis*. FEMS Microbiol Lett 115:27–32

Andres Y, Redercher S, Gerente C, Thouand G (2001) Contribution of biosorption of radionuclides in the environment. J Radioanal Nucl Chem 247:89–93

Antunes SC, Pereira R, Marques SM, Castro BB, Gonçalves F (2011) Impaired microbial activity caused by metal pollution: a field study in a deactivated uranium mining area. Sci Total Environ 410:87–95

Barkay T, Schaefer J (2001) Metal and radionuclide bioremediation: issues, considerations and potentials. Curr Opin Microbiol 4:318–323

Barns SM, Cain EC, Sommerville L, Kuske CR (2007) *Acidobacteria* phylum sequences in uranium-contaminated subsurface sediments greatly expand the known diversity within the phylum. Appl Environ Microbiol 73:3113–3116

Beazley MJ, Martinez RJ, Sobecky PA, Webb SM, Taillefert M (2007) Uranium biomineralization as a result of bacterial phosphatase activity: insights from bacterial isolates from a contaminated subsurface. Environ Sci Technol 41:5701–5707

Beazley MJ, Martinez RJ, Sobecky PA, Webb SM, Taillefert M (2009) Nonreductive biomineralization of uranium(VI) phosphate via microbial phosphatase activity in anaerobic conditions. Geomicrobiol J 26:431–441

Beller HR (2005) Anaerobic, nitrate-dependent oxidation of U(IV) oxide minerals by the chemolithoautotrophic bacterium *Thiobacillus denitrificans*. Appl Environ Microbiol 71:2170–2174

Beyenal H, Sani RK, Peyton BM, Dohnalkova AC, Amonette JE, Lewandowski Z (2004) Uranium immobilization by sulfate-reducing biofilms. Sci Total Environ 38:2067–2074

Bondici VF, Lawrence JR, Khan NH, Hill JE, Yergeau E, Wolfaardt GM, Warner J, Korber DR (2013) Microbial communities in low permeability, high pH uranium mine tailings: characterization and potential effects. J Appl Microbiol 114:1671–1686

Bosecker K (1997) Bioleaching: metal solubilization by microorganisms. FEMS Microbiol Rev 20:591–604

Boswell CD, Dick RE, Macaskie LE (1999) The effect of heavy metals and other environmental conditions on the anaerobic phosphate metabolism of *Acinetobacter johnsonii*. Microbiology 145:1711–1720

Brodie EL, DeSantis TZ, Joyner DC, Baek SM, Larsen JT, Andersen GL, Hazen TC, Richardson PM, Herman DJ, Tokunaga TK, Wan JM, Firestone MK (2006) Application of a high-density oligonucleotide microarray approach to study bacterial population dynamics during uranium reduction and reoxidation. Appl Environ Microbiol 72:6288–6298

Brzoska RM, Bollmann A (2016) The long-term effect of uranium and pH on the community composition of an artificial consortium. FEMS Microbiol Ecol 92(1)

Cardenas E, Wu WM, Leigh MB, Carley J, Carroll S, Gentry, Luo TJ, Watson D, Gu B, Ginder-Vogel M, Kitanidis PK, Jardine PM, Zhou J, Criddle CS, Marsh TL, Tiedje JA (2008) Microbial communities in contaminated sediments, associated with bioremediation of uranium to submicromolar levels. Appl Environ Microbiol 74:3718–3729

Cerdá J, González S, Ríos JM, Quintana T (1993) Uranium concentrates bioproduction in Spain: a case study. FEMS Microbiol Rev 11:253–259

Chang YJ, Peacock AD, Long PE, Stephen JR, McKinley JP, Macnaughton SJ, Hussain AKA, Saxton M, White DC (2001) Diversity and characterization of sulfate-reducing bacteria in groundwater at a uranium mill tailings site. Appl Environ Microbiol 67:3149–3160

Chapon V, Piette L, Berthomieu C, Vesvres MH, Le Marrec C, Coppin L, Février L, Martin-Garin A, Sergean TC (2012) Microbial diversity in contaminated soils along the T22 trench of the Chernobyl experimental platform. Appl Geochem 27:1375–1383

Cho K, Zholi A, Frabutt D, Flood M, Floyd D, Tiquia SM (2012) Linking bacterial diversity and geochemistry of uranium-contaminated groundwater. Environ Technol 33:1629–1640

Chojnacka K (2010) Biosorption and bioaccumulation-the prospects for practical applications. Environ Int 36:299–307

Choudhary S, Sar P (2009) Characterization of a metal resistant *Pseudomonas* sp. isolated from uranium mine for its potential in heavy metal (Ni^{2+}, Co^{2+}, Cu^{2+}, and Cd^{2+}) sequestration. Bioresour Technol 100:2482–2492

Choudhary S, Sar P (2011a) Uranium biomineralization by a metal resistant *Pseudomonas aeruginosa* strain isolated from contaminated mine waste. J Hazard Mater 186:336–343

Choudhary S, Sar P (2011b) Identification and characterization of uranium accumulation potential of a uranium mine isolated *Pseudomonas* strain. W J Microbiol Biotechnol. https://doi.org/10.1007/s11274-010-0637-7

Choudhary S, Sar P (2011c) Identification and characterization of uranium accumulation potential of a uranium mine isolated *Pseudomonas* strain. World J Microbiol Biotechnol 27:1795–1801

Choudhary S, Sar P (2015) Interaction of uranium (VI) with bacteria: potential applications in bioremediation of U contaminated oxic environments. Rev Environ Sci Biotechnol 14:347–355

Choudhary S, Islam E, Kazy SK, Sar P (2012) Uranium and other heavy metal resistance and accumulation in bacteria isolated from uranium mine wastes. J Environ Sci Health Part A 47:622–637

Chourey K, Nissen S, Vishnivetskaya T, Shah M, Pfiffner S, Hettich RL, Löffler FE (2013) Environmental proteomics reveals early microbial community responses to biostimulation at a uranium- and nitrate-contaminated site. Proteomics 13(18–19):2921–2930

Coates JD, Bhupathiraju VK, Achenbach LA, McInerney MJ, Lovley DR (2001) *Geobacter hydrogenophilus* and *Geobacter chapellei* and *Geobacter grbiciae*, three new, strictly anaerobic, dissimilatory Fe(III)-reducers. Int J Syst Evol Microbiol 51:581–588

Dhal PK, Sar P (2014) Microbial communities in uranium mine tailings and mine water sediment from Jaduguda U mine, India: A culture independent analysis. J Environ Sci Health Part A 49:694–709

Dhal PK, Islam E, Kazy SK, Sar P (2011) Culture independent molecular analysis of bacterial diversity in uranium -ore/-mine waste contaminated and non-contaminated sites from uranium mines. 3 Biotech 1:261–272

Douglas S, Beveridge TJ (1998) Mineral formation by bacteria in natural microbial communities. FEMS Microbiol Rev 26:79–88

Elias DA, Krumholz LR, Wong D, Long PE, Suflita JM (2003) Characterization of microbial activities and U reduction in a shallow aquifer contaminated by uranium mill tailings. Microb Ecol 46:83–91

Fields MW, Yan T, Rhee SK, Carroll SL, Jardine PM, Watson DB, Criddle CS, Zhou J (2005) Impacts on microbial communities and cultivable isolates from groundwater contaminated with high levels of nitric acid-uranium waste. FEMS Microbiol Ecol 53:417–428

Fletcher KE, Boyanov MI, Thomas SH, Wu Q, Kemner KM, Löffler FE (2010) U(VI) reduction to mononuclear U(IV) by desulfitobacterium species. Environ Sci Technol 44:4705–4709

Fowle DA, Fein JB, Martin AM (2000) Experimental study of uranyl adsorption onto *Bacillus subtilis*. Environ Sci Technol 34:3737–3741

Francis AJ, Gillow JB, Dodge CJ, Harris R, Harris R, Beveridge TJ, Papenguth HW (2004) Uranium association with halophilic and non-halophilic bacteria and archaea. Radiochim Acta 92:481–488

Francis AJ, Dodge CJ, Gillow JB (2008) Reductive Dissolution of Pu(IV) by *Clostridium* sp. under anaerobic conditions. Environ Sci Technol 42:2355–2360

Fredrickson JK, Zachara JM, Balkwill DL, Kennedy D, Li SM, Kostandarithes HM, Daly MJ, Romine MF, Brockman FJ (2004) Geomicrobiology of high-level nuclear waste-contaminated vadose sediments at the hanford site, Washington state. Appl Environ Microbiol 70:4230–4241

Fuchs T, Huber H, Teiner K, Burggraf S, Stetter KO (1995) *Metallosphaera prunae*, sp. nov., a novel metal-mobilizing, isolated from a uranium mine in Germany. Syst Appl Microbiol 18:560–566

Gadd GM (1992) Metals and microorganisms: a problem of definition. FEMS Microbiol Lett 79:197–203

Gadd GM (2004) Microbial influence on metal mobility and application for bioremediation. Geoderma 122:109–119

Gadd GM (2010) Metals, minerals and microbes: geomicrobiology and bioremediation. Microbiology 156:609–643

Geissler A (2007) Prokaryotic microorganisms in uranium mining waste piles and their interactions with uranium and other heavy metals. PhD Thesis, TU Bergakademie Freiberg, Freiberg

Geissler A, Selenska-Pobell S (2005) Addition of U(VI) to a uranium mining waste sample and resulting changes in the indigenous bacterial community. Geobiology 3:275–285

Goodman AE, Khalid AM, Ralph BJ (1981) Microbial ecology of Rum Jungle I; environmental study of sulphidic overburden dumps, experimental heap leach piles and tailings dam area. Australian Atomic Energy Commission, Report AAEC/E531, Australia

Gorby YA, Lovley DR (1992) Enzymatic uranium precipitation. Environ Sci Technol 26:205–207

Gorman-Lewis D, Elias PE, Fein JB (2005) Adsorption of aqueous uranyl complexes onto *Bacillus subtilis* cells. Environ Sci Technol 39:4906–4912

Green SJ, Prakash O, Jasrotia P, Overholt WA, Cardenas E, Hubbard D, Tiedje JM, Watson DB, Schadt CW, Brooks SC, Kostka JE (2012) Denitrifying bacteria from the genus *Rhodanobacter* dominate bacterial communities in the highly contaminated subsurface of a nuclear legacy waste site. Appl Environ Microbiol 78:1039–1047

Hafez M, Ibrahim M, Abdel-Razek A, Abu-Shady M (2002) Biosorption of some ions on different bacterial species from aqueous and radioactive waste solutions. J Radioanal Nucl Chem 252:179–185

Hemme CL, Deng Y, Gentry TJ, Fields MW, Wu L, Barua S, Barry K, Tringe SG, Watson DB, He Z, Hazen TC, Tiedje JM, Rubin EM, Zhou J (2010) Metagenomic insights into evolution of a heavy metal-contaminated groundwater microbial community. ISME J 4:660–672

Hwang C, Wu W, Gentry TJ, Carley J, Corbin GA, Carroll SL, Watson DB, Jardine PM, Zhou J, Criddle CS, Fields MW (2009) Bacterial community succession during in situ uranium bioremediation: spatial similarities along controlled flow paths. ISME J 3:47–64

Islam E, Sar P (2011) Culture-dependent and -independent molecular analysis of the bacterial community within uranium ore. J Basic Microbiol 4:372–384

Islam E, Sar P (2016) Diversity, metal resistance and uranium sequestration abilities of bacteria from uranium ore deposit in deep earth stratum. Ecotoxicol Environ Saf 127:12–21

Islam E, Dhal PK, Sufia KK, Sar P (2011) Molecular analysis of bacterial communities in uranium ores and surrounding soils from Banduhurang open cast uranium mine, India: a comparative study. J Environ Sci Health Part A 46:271–280

Ivanova IA, Stephen JR, Chang Y-J, Brüggemann J, Long PE, McKinley JP, Kowalschuk GA, White DC, Macnaughton SJ (2000) A survey of 16S rRNA and *amoA* genes related to autotrophic ammonia-oxidizing bacteria of the β-subdivision of the class *proteobacteria* in contaminated groundwater. Can J Microbiol 46:1012–1020

Jroundi F, Merroun ML, Arias JM, Rossberg A, Selenska-Pobell S, González-Muñoz MT (2007) Spectroscopic and microscopic characterization of uranium biomineralization in *Myxococcus xanthus*. Geomicrobiol J 24:441–449

Kashefi K, Lovely DR (2000) Reduction of Fe(III), Mn(IV), and toxic metals at 100 °C by *Pyrobaculum islandicum*. Appl Environ Microbiol 66:1050–1056

Kazy SK, D'Souza SF, Sar P (2009) Uranium and thorium sequestration by a *Pseudomonas* sp.: mechanism and chemical characterization. J Hazard Mater 163:65–72

Kelly SD, Kemner KM, Fein JB, Fowle DA, Boyanov MI, Bunker BA, Yee N (2002) X-ray absorption fine structure determination of pH dependent U-bacterial cell wall interactions. Geochim Cosmochim Acta 65:3855–3871

Kenarova A, Radeva G, Traykov I, Boteva S (2014) Community level physiological profiles of bacterial communities inhabiting uranium mining impacted sites. Ecotox Environ Safe 100:226–232

Khan NH, Bondici VF, Medihala PG, Lawrence JR, Wolfaardt GM, Warner J, Korber DR (2013) Bacterial diversity and composition of an alkaline uranium mine tailings-water interface. J Microbiol 51:558–569

Kieft TL, Fredrickson JK, Onstott TC, Gorby YA, Kostandarithes HM, Bailey TJ, Kennedy DW, Li SW, Plymale AE, Spadoni CM, Gray MS (1999) Dissimilatory reduction of Fe(III) and other electron acceptors by a *Thermus* isolate. Appl Environ Microbiol 65:1214–1421

Knopp R, Panak PJ, Wray LA, Renninger NS, Keasling JD, Nitsche H (2003) Laser spectroscopic studies of interactions of UVI with bacterial phosphate species. Chem Eur J 9:2812–2818

Konopka A, Plymale AE, Carvajal DA, Lin XJ, Mckinley JP (2013) Environmental controls on the activity of aquifer microbial communities in the 300 Area of the Hanford site. Microb Ecol 66:889–896

Kotelnikova S, Macario AJL, Pedersen K (1998) *Methanobacterium subterraneum* sp. nov., a new alkaliphilic, eurythermic and halotolerant methanogen isolated from deep granitic groundwater. Int J Syst Evol Microbiol 48:357–367

Krueger S, Olson GJ, Johnsonbaugh D, Beveridge TJ (1993) Characterization of the binding of gallium, platinum, and uranium to *Pseudomonas fluorescens* by small-angle X-Ray scattering and transmission electron microscopy. Appl Environ Microbiol 59:4056–4064

Kumar R, Nongkhlaw M, Acharya C, Joshi SR (2013) Uranium (U)-tolerant bacterial diversity from U ore deposits of Domiasiat in North-East India and their prospective utilization in bioremediation. Microbes Environ 28:33–41

Lack J, Chaudhuri S, Kelly S, Kemner K, O´Connor S, Coates J (2002) Immobilization of radionuclides and heavy metals through anaerobic bio-oxidation of Fe(II). Appl Environ Microbiol 68:2704–2710

Landa ER (2005) In: Laskin AI, Bennett JW, Gadd GM (eds) Microbial biogeochemistry of uranium mill tailings advances in applied microbiology, vol 57. Academic Press, Burlington, pp 113–130

Langley S, Beveridge TJ (1999) Effect of O-side-chain lipopolysaccharide chemistry on metal binding. Appl Environ Microbiol 65:489–498

Leigh MB, Wu W, Cardenas E, Uhlik O, Carroll S, Gentry T, Marsh TL, Zhou J, Jardine P, Criddle CS, Tiedje JM (2014) Microbial communities biostimulated by ethanol during uranium (VI) bioremediation in contaminated sediment as shown by stable isotope probing. Front Environ Sci Eng 2015(9):453–464

Lloyd J, Gadd G (2011) The geomicrobiology of radionuclides. Geomicrobiol J 28:383–386

Lloyd JR, Renshaw JC (2005) Microbial transformations of radionuclides: fundamental mechanisms and biogeochemical implication. Met Ions Biol Syst 44:205–240

Lovley D (1993) Dissimilatory metal reduction. Annu Rev Microbiol 47:263–290

Lovley DR (2003) Cleaning up with genomics: applying molecular biology to bioremediation. Nat Rev Microbiol 1:35–44

Lovley DR, Coates JD (1997) Bioremediation of metal contamination. Curr Opin Biotechnol 8:285–289

Lovley DR, Phillips EJP (1992) Reduction of uranium by *Desulfovibrio desulfuricans*. Appl Environ Microbiol 58:850–856

Lovley DR, Philips EJP, Gorby YA, Landa ER (1991) Microbial reduction of uranium. Nature 350:413–416

Maleke M, Williams P, Castillo J, Botes E, Ojo A, DeFlaun M, van Heerden E (2014) Optimization of a bioremediation system of soluble uranium based on the biostimulation of an indigenous bacterial community. Environ Sci Pollut Res 22:8442–8450

Malekzadeh F, Farazmand A, Ghafourtian H, Shahamat M, Levin M, Colwell RR (2002) Uranium accumulation by a bacterium isolated from electroplating effluent. World J Microbiol Biotechnol 18:295–302

Marsili E, Beyenal H, Di Palma L, Merli C, Dohnalkova A, Amonette JE, Lewandowski Z (2007) Uranium immobilization by sulfate-reducing biofilms grown on hematite, dolomite, and calcite. Environ Sci Technol 41:8349–8354

Martinez RJ, Wang Y, Raimondo MA, Coombs JM, Barkay T, Sobecky PA (2006) Horizontal gene transfer of PIB-type ATPases among bacteria isolated from radionuclide-and metal-contaminated subsurface soils. Appl Environ Microbiol 72:3111–3118

Martinez RJ, Beazley MJ, Taillefert M, Arakaki AK, Skolnick J, Sobecky PA (2007) Aerobic uranium (VI) bioprecipitation by metal-resistant bacteria isolated from radionuclide-and metal-contaminated subsurface soils. Environ Microbiol 9:3122–3133

McLean J, Beveridge TJ (2001) Chromate reduction by a *Pseudomonad* isolated from a site contaminated with chromated copper arsenate. Appl Environ Microbiol 67:1076–1084

Merroun ML, Selenska-Pobell S (2001) Interactions of three eco-types of *Acidithiobacillus ferrooxidans* with U(VI). Biometals 14:171–179

Merroun ML, Selenska-Pobell S (2008) Bacterial interactions with uranium: an environmental perspective. J Contam Hydrol 102:285–229

Merroun ML, Hennig C, Rossberg A, Geipel G, Reich T, Selenska-Pobell S (2002) Molecular and atomic analysis of the uranium complexes formed by three ecotypes of *Acidithiobacillus ferrooxidans*. Biochem Sc Trans 30:669–672

Merroun ML, Hennig C, Rossberg A, Reich T, Selenska-Pobell S (2003) Characterization of U(VI)-*Acidithiobacillus ferrooxidans* complexes using EXAFS, transmission electron microscopy, and energy-dispersive X-ray analysis. Radiochim Acta 91:583–592

Merroun ML, Raff J, Rossberg A, Hennig C, Reich T, Selenska-Pobell S (2005) Complexation of uranium by cells and S-layer sheets of *Bacillus sphaericus* JG-A12. Appl Environ Microbiol 71:5532–5543

Michalsen MM, Peacock AD, Spain AM, Smithgal AN, White DC, Sanchez-Rosario Y, Krumholz LR, Istok JD (2007) Changes in microbial community composition and geochemistry during uranium and technetium bioimmobilization. Appl Environ Microbiol 73:5885–5896

Miller CL, Landa ER, Updegraff DM (1987) Ecological aspects of microorganisms inhabiting uranium mill tailings. Microb Ecol 14:141–155

Mondani L, Benzerara K, Carrière M, Christen R, Mamindy-Pajany Y, Février L, Marmier N, Achouak W, Nardoux P, Berthomieu C, Chapon V (2011) Influence of uranium on bacterial communities: a comparison of natural uranium-rich soils with controls. PLoS One 6:e25771

Mosher JJ, Phelps TJ, Podar M, Hurt RA Jr, Campbell JH, Drake MM, Moberly JG, Schadt CW, Brown SD, Hazen TC, Arkin AP, Palumbo AV, Faybishenko BA, Elias DA (2012) Microbial community succession during lactate amendment and electron acceptor limitation reveals a predominance of metal-reducing *Pelosinus* spp. Appl Environ Microbiol 78:2082–2091

Nakajima A, Tsuruta T (2004) Competitive biosorption of thorium and uranium by *Micrococcus luteus*. J Radioanal Nucl Chem 260:13–18

Nedelkova M, Merroun ML, Rossberg A, Hennig C, Selenska-Pobell S (2007) *Microbacterium* isolates from the vicinity of a radioactive waste depository and their interactions with uranium. FEMS Microbiol Ecol 59:694–705

Nilgiriwala KS, Alahari A, Rao AS, Apte SK (2008) Cloning and over expression of alkaline phosphatase PhoK from *Sphingomonas* sp. strain BSAR-1 for bioprecipitation of uranium from alkaline solutions. Appl Environ Microbiol 74:5516–5523

Nostrand JDV, Wu L, Wu WM, Huang Z, Gentry TJ, Deng Y, Carley J, Carroll S, He Z, Gu B, Luo J, Criddle CS, Watson DB, Jardine PM, Marsh TL, Tiedje JM, Hazen TC, Zhou J (2011)

Dynamics of microbial community composition and function during in situ bioremediation of a uranium contaminated aquifer. Appl Environ Microbiol 77:3860–3869

OECD (2008) Consensus document on information used in the assessment of environmental applications involving Acinetobacter, Series on harmonization of regulatory oversight in biotechnology No. 46. Organisation for Economic Co-operation and Development (OECD), Paris

Panak P, Hard BC, Pietzsch K, Kutschke S, Röske K, Selenska-Pobell S, Bernhard G, Nitsche H (1998) Bacteria from uranium mining waste pile: interactions with U(VI). J Alloys Compd 271–273:1

Panak PJ, Knopp R, Booth CH, Nitsche H (2002) Spectroscopic studies on the interaction of U (VI) with *Bacillus sphaericus*. Radiochim Acta 90:779–783

Pereira SG, Albuquerque L, Nobre MF, Tiago I, Veríssimo A, Pereira A, da Costa MS (2012) *Pullulanibacillus uraniitolerans* sp. nov., a new acidophilic, U(VI)-resistant species isolated from an acid uranium mill tailing effluent and emended description of the genus *Pullulanibacillus*. Int J Syst Evol Microbiol. https://doi.org/10.1099/ijs.0.040923-0

Petrie L, North NN, Dollhopf SL, Balkwill DL, Kostka JE (2003) Enumeration and characterization of iron(III)-reducing microbial communities from acidic subsurface sediments contaminated with uranium(VI). Appl Environ Microbiol 69:7467–7479

Pollmann K, Raff J, Merroun M, Fahmy K, Selenska-Pobell S (2005) Metal binding by bacteria from uranium mining waste piles and its technological applications. Biotechnol Adv 24:58–68

Porat I, Vishnivetskaya TA, Mosher JJ, Brandt CC, Yang ZK, Brooks SC, Liang L, Drake MM, Podar M, Brown SD, Palumbo AV (2010) Characterization of archaeal community in contaminated and uncontaminated surface stream sediments. Microb Ecol 60:784–795

Prakash O, Gihring TM, Dalton DD, Chin KJ, Green SJ, Akob DM, Wanger G, Kostka JE (2010) *Geobacter daltonii* sp. nov., an Fe(III)- and uranium(VI)-reducing bacterium isolated from a shallow subsurface exposed to mixed heavy metal and hydrocarbon contamination. Int J Syst Evol Microbiol 3:546–553

Radeva G, Selenska-Pobell S (2005) Bacterial diversity in water samples from uranium wastes as demonstrated by 16S rDNA and ribosomal intergenic spacer amplification retrievals. Can J Microbiol 51:910–923

Radeva G, Kenarova A, Bachvarova V, Flemming K, Popov I, Vassilev D, Selenska-Pobell S (2013) Bacterial diversity at abandoned uranium mining and milling sites in Bulgaria as revealed by 16S rRNA genetic diversity study. Water Air Soil Pollut 224:1–14

Radeva G, Kenarova A, Bachvarova V, Flemming K, Popov I, Vassilev D, Selenska-Pobell S (2014) Phylogenetic diversity of archaea and the Archaeal ammonia Monooxygenase gene in uranium mining-impacted locations in Bulgaria. Archaea 2014:1–10

Ragon M, Restoux G, Moreira D, Møller AP, López-García P (2011) Sunlight-exposed biofilm microbial communities are naturally resistant to chernobyl ionizing-radiation levels. PLoS One 6:e21764

Rastogi G, Osman S, Kukkadapu R, Engelhard M, Vaishampayan P, Andersen G, Sani R (2010a) Microbial and mineralogical characterizations of soils collected from the deep biosphere of the former Homestake Gold Mine, South Dakota. Microb Ecol 60:539–550

Rastogi G, Osman S, Vaishampayan P, Andersen G, Stetler L, Sani R (2010b) Microbial diversity in uranium mining-impacted soils as revealed by high-density 16S microarray and clone library. Microb Ecol 59:94–108

Reardon CL, Cummings DE, Petzke LM, Kinsall BL, Watson DB, Peyton BM, Geesey GG (2004) Composition and diversity of microbial communities recovered from surrogate minerals incubated in an acidic uranium-contaminated aquifer. Appl Environ Microbiol 70:6037–6046

Rittmann BE, Hausner M, Löffler F, Love NG, Muyzer G, Okabe S, Oerther DB, Peccia J, Raskin L, Wagner M (2006) A vista for microbial ecology and environmental biotechnology. Environ Sci Technol 40:1096–1103

Sánchez-Castro I, Amador-García A, Moreno-Romero C, López-Fernández M, Phrommavanh V, Nos J, Descostes M, Merroun ML (2017) Screening of bacterial strains isolated from uranium mill tailings porewaters for bioremediation purposes. J Environ Radioact 166:130–141

Satchanska G, Selenska-Pobell S (2005) Bacterial diversity in the uranium mill-tailings Gittersee as estimated via a 16S rDNA approach. C R Acad Bulg Sci 58:1105–1112

Satchanska G, Golovinsky E, Selenska-Pobell S (2004) Bacterial diversity in a soil sample from uranium mining waste pile as estimated via a culture-independent 16S rDNA approach. C R Acad Bulg Sci 57:75–82

Schippers A, Hallmann R, Wentzien S, Sand W (1995) Microbial diversity in uranium mine waste heaps. Appl Environ Microbiol 61:2930–2935

Selenska-Pobell S (2002) Diversity and activity of bacteria in uranium waste piles. In: Keith-Roach M, Livens F (eds) Interactions of microorganisms with radionuclides. Elsevier, Oxford, pp 225–253

Selenska-Pobell S, Panak P, Miteva V, Boudakov I, Bernhard G, Nitsche H (1999) Selective accumulation of heavy metals by three indigenous *Bacillus* strains, *B. cereus, B. megaterium* and *B. sphaericus,* from drain waters of a uranium waste pile. FEMS Microbiol Ecol 29:59–67

Selenska-Pobell S, Kampf G, Hemming K, Radeva G, Satchanska G (2001) Bacterial diversity in soil samples from two uranium waste piles as determined by rep-APD, RISA and 16S rDNA retrieval. Anton Van Leeuw 79:149–161

Selenska-Pobell S, Flemming K, Tzvetkova T, Raff J, Schnorpfeil M, Geißler A (2002) Bacterial communities in uranium mining waste piles and their interaction with heavy metals. In: Merkel BJ, Planer-Friedrich B, Wolkersdorfer C (eds) Uranium in the aquatic environment. Springer, Berlin

Shelobolina ES, Sullivan SA, O'Neill KR, Nevin KP, Lovley DR (2004) Isolation, characterization, and U(VI)-reducing potential of a facultatively anaerobic, acid-resistant bacterium from low-pH, nitrate- and U(VI)-contaminated subsurface sediment and description of *Salmonella subterranea* sp. nov. Appl Environ Microbiol 70:2959–2965

Silver M (1987) Distribution of iron-oxidizing bacteria in the nordic uranium tailings deposit, Elliot Lake, Ontario, Canada. Appl Environ Microbiol 53:846–852

Sitte J, Löffler S, Burkhardt EM, Goldfarb KC, Büchel G, Hazen TC, Küsel K (2015) Metals other than uranium affected microbial community composition in a historical uranium-mining site. Environ Sci Pollut Res 22:19326

Sivaswamy V, Boyanov MI, Peyton BM, Viamajala S, Gerlach R, Apel WA, Sani RK, Dohnalkova A, Kemner KM, Borch T (2011) Multiple mechanisms of uranium immobilization by *Cellulomonas* sp. strain ES6. Biotechnol Bioeng 108:264–276

Stein LY, Jones G, Alexander B, Elmund K, Wright-Jones C, Nealson KH (2002) Intriguing microbial diversity associated with metal-rich particles from a freshwater reservoir. FEMS Microbiol Ecol 42:431–440

Strandberg GW, Shumate SE, Parrott JR (1981) Microbial cells as biosorbents for heavy metals: accumulation of uranium by *Saccharomyces cerevisiae* and *Pseudomonas aeruginosa*. Appl Environ Microbiol 41:237–245

Suzuki SY, Banfield JF (1999) Geomicrobiology of uranium. In: Burns PC, Finch R (eds) Uranium: mineralogy, geochemistry and the environment, vol 38. Mineralogical Society of America, Washington, DC, pp 393–432

Suzuki SY, Banfield JF (2004) Resistance to, and accumulation of, uranium by bacteria from a uranium-contaminated. Geomicrobiol J 21:113–121

Suzuki SY, Suko T (2006) Geomicrobiological factors that control uranium mobility in the environment: Update on recent advances in the bioremediation of uranium-contaminated sites. J Mineral Petrol Sci 101:299–307

Suzuki Y, Kelly SD, Kemner KM, Banfield JF (2005) Direct microbial reduction and subsequent preservation of uranium in natural nearsurface sediment. Appl Environ Microbiol 71:1790–1797

Tabak HH, Lens P, van Hullebusch ED, Dejonghe W (2005) Developments in bioremediation of soils and sediments polluted with metals and radionuclides – 1. Microbial processes and mechanisms affecting bioremediation of metal contamination and influencing metal toxicity and transport. Rev Environ Sci Biotechnol 4:115–156

Takai K, Moser DP, DeFlaun M, Onstott TC, Fredrickson JK (2001) Archaeal diversity in waters from deep South African gold mines. Appl Environ Microbiol 67:5750–5760

Templeton A, Knowles E (2009) Microbial transformations of minerals and metals: recent advances in geomicrobiology derived from synchrotron-based X-ray spectroscopy and X-ray microscopy. Annu Rev Earth Planet Sci 37:391

Tsezos M, Volesky B (1981) Biosorption of uranium and thorium. Biotechnol Bioeng 23:583–604

Tsezos M, Remoudaki E, Angelatou V (1997) Biosorption sites of selected metals using electron microscopy. Comp Biochem Physiol 118A:481–487

Tsuruta T (2007) Removal and recovery of uranium using microorganisms isolated from North American uranium deposits. Am J EnvironSci 3:60–66

Waldron PJ, Wu L, Van Nostrand JD, Schadt CW, He Z, Watson DB, Jardine PM, Palumbo AV, Hazen TC, Zhou J (2009) Functional gene array-based analysis of microbial community structure in groundwaters with a gradient of contaminant levels. Environ Sci Technol 43:3529–3534

Wolfaardt GM, Hendry MJ, Korber DR (2008) Microbial distribution and diversity in saturated, high pH, uranium mine tailings, Saskatchewan, Canada. Can J Microbiol 54:932–940

Wu WM, Carley J, Gentry T, Ginder-Vogel MA, Fienen M, Mehlhorn T, Yan H, Caroll S, Pace MN, Nyman J, Luo J, Gentile ME, Fields MW, Hickey RF, Gu B, Watson D, Cirpka OA, Zhou J, Fendorf S, Kitanidis PK, Jardine PM, Criddle CS (2006) Pilot-scale in situ bioremedation of uranium in a highly contaminated aquifer. 2. Reduction of U(VI) and geochemical control of U(VI) bioavailability. Environ Sci Technol 40:3986–3995

Yan X, Luo X (2015) Radionuclides distribution, properties, and microbial diversity of soils in uranium mill tailings from southeastern China. J Environ Radioact 139:85–90

Yi Z, Lian B (2012) Adsorption of U(VI) by *Bacillus mucilaginosus*. J Radioanal Nucl Chem 293:321–329

Yong P, Macaskie LE (1998) Bioaccumulation of lanthanum, uranium and thorium, and use of a model system to develop a method for the biologically-mediated removal of plutonium from solution. J Chem Technol Biotechnol 71:15–26

Zhang P, Wu W-M, Van Nostrand J, Deng Y, He Z, Gihring T et al (2015) Dynamic succession of groundwater functional microbial communities in response to emulsified vegetable oil amendment during sustained in situ U(VI) reduction. Appl Environ Microbiol 81:4164–4172

Yan X, Luo X, Zhao M (2016) Metagenomic analysis of microbial community in uranium-contaminated soil. Appl Microbiol Biotechnol 100(1):299–310

Microbial Cycling of Greenhouse Gases and Their Impact on Climate Change

7

Bharati Kollah, Ashok Kumar Patra, and Santosh Ranjan Mohanty

Abstract

Greenhouse gas (GHG) emission from agriculture contributes significantly to the global climate change. The major greenhouse gases emitted from agriculture are methane (CH_4) and nitrous oxide (N_2O). These two greenhouse gases have higher global warming potential than carbon dioxide (CO_2). The manuscript embodies biogeochemical cycling of CH_4 and N_2O. Microbial pathways of methanogenesis, CH_4 oxidation, nitrification, and denitrification are outlined. Information on the agricultural strategies to mitigate GHG emission from soils are discussed. The review highlights significance of low-affinity methanotrophs that can be activated by repeated enrichment of high CH_4 concentration, as global climate regulators. Iron redox cycling is also linked with soil CH_4 uptake as repeated Fe^{3+} reduction and Fe^{2+} oxidation decline crystalline Fe fraction that enhances CH_4 consumption by stimulating pmoA gene of methanotrophs. Studies suggested alternate flooding and drying as a potential approach to mediate atmospheric CH_4 uptake in flooded soil. N_2O emission from soil is the outcome of both nitrification and denitrification. However, in upland soil N_2O emission occurs through nitrification and through denitrification from flooded soil ecosystem. Nitrogen-fixing *Rhizobium* sp. also produces N_2O, and these bacteria can be manipulated to mitigate N_2O emission by activating N_2O reductase (*nosZ* gene). It is concluded that apart from regular agricultural resource management strategies, there is need of genetically manipulated soil microorganisms to effectively mitigate climate change.

Keywords

Greenhouse gas · CH_4 · N_2O · Microbes · Climate change

B. Kollah · A. K. Patra · S. R. Mohanty (✉)
ICAR-Indian Institute of Soil Science (IISS), Bhopal, India
e-mail: mohantywisc@gmail.com

T. K. Adhya et al. (eds.), *Advances in Soil Microbiology: Recent Trends and Future Prospects*, Microorganisms for Sustainability 3,
https://doi.org/10.1007/978-981-10-6178-3_7

7.1 Introduction

Climate change has adversely affected agricultural production and is likely to continue to do so during the coming decades. The impact of climate change is anticipated to be increasingly negative on most crops and livestock especially in the tropics. Apart from direct phonological and physiological impacts, there will be further decline in crop and livestock production due to additional stresses like weeds, diseases, and insect pests. Increase in the atmospheric concentrations of greenhouse gases (GHGs) is considered to be the main cause of climate change. The three major GHGs are CO_2, CH_4, and N_2O. However, CH_4 and N_2O are more important greenhouse gases than CO_2 because of their global warming potential. The ability of CH_4 and N_2O molecules to absorb infrared radiation makes these gases 20–30 and 200–300, respectively, more effective than CO_2 as a greenhouse gas, resulting in significant contribution to the radiative forcing of the atmosphere and associated global climate change. Atmospheric CO_2 level increased at the rate of 1.5 ppm yr^{-1} during 1980–1990, but in the last decade, CO_2 concentration increased at the rate of 2.0 ppm yr^{-1} (Raupach et al. 2007). On the contrary, other potential greenhouse gas CH_4 increased at the rate of 11.5 ppb yr^{-1} during 2014–2015 and at the rate of 5.7 ± 1.2 ppb yr^{-1} during 2007–2013. The most potent greenhouse gas, N_2O, has increased by 18% than the preindustrial period. Its concentration is linearly increasing at the rate of 0.26% yr^{-1} since the last few decades (Forster et al. 2007).

Being involved in carbon, nitrogen, phosphorous, and other elemental cycles, microorganisms are also responsible for both the production and consumption of greenhouse gases. Soil microbes thus exhibit both positive and negative feedback to atmospheric GHGs. These microbial activities are also influenced by the climatic factors. The reason for this microbial response to climate change is due to the complex interaction of microbes with higher organisms and environment. This complex interaction is not clearly known making it difficult to accurately predict the response of microbes to climate change. It has been noted that human activities have increased the production of greenhouse gases originating from microbial community. The chapter aims to provide information on (1) the microbial processes regulating cycling of the CH_4 and N_2O in soil and (2) microbial strategies to mitigate GHG emission from agricultural soils.

7.2 Biogeochemical Cycling of CH_4

Wetlands and rice soils are the large source of CH_4. It is the final product of anaerobic methanogenic respiration. The production of CH_4 by anaerobic methanogens includes reduction of methanol, CO_2 and cleavage of acetate, as well as biosynthesis of methylated compounds (Angel et al. 2012). Methanogenic archaea are two types: acetoclastic and hydrogenotrophic. These two groups of methanogens play a vital role for all biogenically produced CH_4 in anoxic habitats. Acetate is used as C source by the acetoclastic methanogenic archaea. These groups are represented as Methanosarcinaceae and Methanosaetaceae. Species belonging to

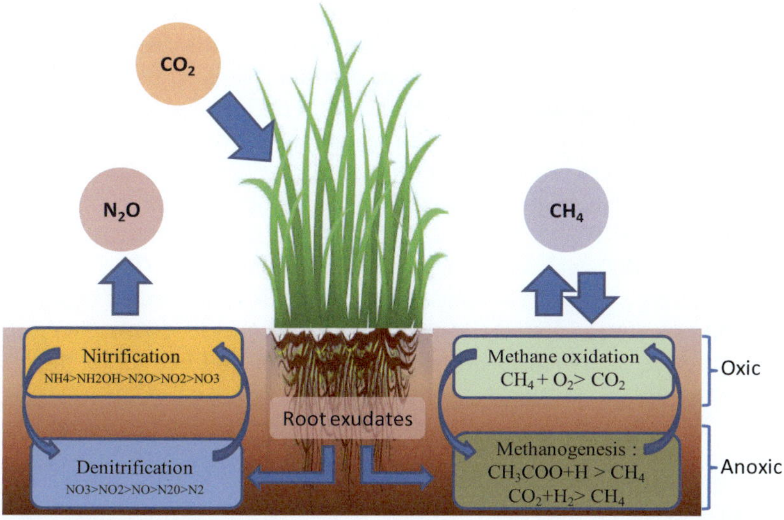

Fig. 7.1 Biogeochemical cycling of methane (CH_4) and nitrous oxide (N_2O) in oxic and anoxic layers of wetland soil and sediments

Methanosaetaceae are abundant in paddy soil when acetate concentration is low, while Methanosarcinaceae are dominant at higher acetate concentration (Eller et al. 2005). Hydrogenotrophic methanogens use H_2 and CO_2 for methanogenesis. The hydrogenotrophic methanogens are Methanocellales, Methanomicrobiales, Methanosarcinales, and Methanobacteriales. In wetlands CH_4 formation from H_2/CO_2 is much larger (up to 67%) than from acetate (33%) (Conrad 1999).

Wetlands are the largest source of CH_4, which contributes one-third of the global CH_4 budget. Methane is produced under anaerobic conditions by the methanogens (Fig. 7.1), but the net amount of CH_4 reaching the atmosphere is influenced by many abiotic factors including soil temperature, pH, nutrient content, and moisture. Plants regulate the flux of CH_4 from wetlands by different processes. Plants can stimulate CH_4 emissions by providing carbon substrates to the methanogens. These C compounds originate from plants as root exudates. Plants also help in the transport of CH_4 from soil to atmosphere by acting as a conduit. Plants create oxidized condition in the rhizosphere that can influence CH_4 oxidation. The relative significance of these processes varies among plant species. Many studies have revealed that CH_4 fluxes can be increased by the presence of vascular plants, while others have found that there can be a decline in CH_4 production.

Rice paddies are the major source of CH_4. In general rice straw is incorporated into the soil during field preparation for cultivating rice. Amendment of rice straw to flooded paddy soil improves soil structure and soil organic carbon in the long term (Zhang et al. 2013). However, such practice potentially increases CH_4 emission from soil into the atmosphere (Yuan et al. 2014b). The decomposed organic matter acts as an electron source to reduce O_2, NO_3, Fe^{3+}, Mn^{4+}, SO_4^{2-}, and CO_2 sequentially in the anaerobic soils. The presence of electron acceptors other than CO_2, i.e., O_2, NO_3, Fe^{3+}, Mn^{4+}, and SO_4^{2-} inhibits methanogenesis (Rissanen et al. 2016). Methanogens respond differently to the incorporated organic residues. The abundance of the methanogenic communities increases during anoxic decomposition of rice straw (Yuan et al. 2014a). It has been reported that rice straw incorporation into soil selectively enhances population of Methanosarcinaceae and Methanobacteriales and decreases methanogens belonging to rice cluster I (RC-I) and Methanomicrobiales (Hernández et al. 2017). Degrading rice straw is colonized by Methanosarcina, Methanobacterium, and RC-I methanogens (Bao et al. 2014). Several pioneering studies have focused on the effects of rice straw application on CH_4 emission in rice paddy soil (Dong et al. 2013; Han et al. 2016). The pattern of CH_4 production is, however, mainly dependent on the soil but not on the types of straw added.

7.2.1 Methane Oxidation

Upland soils are generally well-drained and aerated (oxic) in nature. These soils have major role in the global CH_4 budget as they act as sink for atmospheric CH_4. It is estimated that globally CH_4 consumption is about 30 Tg yr^{-1} (Rice et al. 2016). Although CH_4 consumption occurs in a wide variety of upland soils, the pristine forest soils have been identified as the most promising sinks for atmospheric CH_4 (Lohila et al. 2016; Ťupek et al. 2014). Conversion of pristine forest land to agricultural land can lower the CH_4 uptake capacity (Knox et al. 2015; Tate 2015). Various agricultural factors regulate CH_4 oxidation. Some of these factors are soil compaction, pH and fertilizer application (Ball 2013), and abandonment of agricultural land, or even converting it to forest can potentially increase the atmospheric CH_4 uptake to some extent.

Methanotroph diversity and activity has been studied in different upland soils (Knief et al. 2003; Vanitchung et al. 2014). The diversity of CH_4-oxidizing bacteria is typically assessed by exploring *pmoA* gene. This gene encodes β-subunit of methane monooxygenase (pMMO) enzyme (Mau et al. 2013). Most of the uncultivated methanotrophs are characterized by *pmoA* gene sequences. Similarly, methanotrophs can be identified by analyzing their phospholipid fatty acids (Steger et al. 2015). Methanotrophs are aerobic, gram-negative bacteria and use CH_4 as their sole source of energy. The methanotrophs also degrade various environmental contaminants and are used in various environmental remediation projects. Methanotrophs are also used as source of single-cell protein. Based on the physiological and biochemical characteristics of these microbial groups, the cultured members of the methanotrophs are divided into three groups: type I, type II, and type X. Type I are the members of the class γ-proteobacteria (*Methylomonas*, *Methylococcus*, *Methylomicrobium*, *Methylothermus*, *Methylohalobium*, *Methylocaldum*, and *Methylobacter*). Type II belongs to the class α-proteobacteria (*Methylosinus*, *Methylocella*, *Methylocapsa*, and *Methylocystis*). Type X

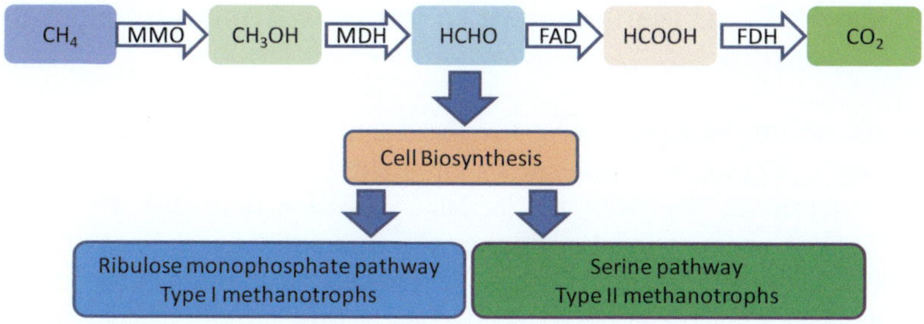

Fig. 7.2 Schematic illustration of CH_4 oxidation pathway carried out by methanotrophs in soil. The enzyme complexes are methane monooxygenase (MMO), methanol dehydrogenase (MDH), formaldehyde dehydrogenase (FAD), and formic acid dehydrogenase (FDH)

methanotrophs belong to the class γ-proteobacteria (*Methylococcus*) and feature characteristic of both type I and II.

Based on the *pmoA* gene, it was revealed that the USCα pmoA clade is commonly found in upland soils (Knief et al. 2006). These belong to *a*-proteobacteria and are related to *Methylocapsa* (Ricke et al. 2004). The USCγ pmoA clade methanotrophs are prevalent in upland soils of neutral or somewhat alkaline pH. Methanotrophs from clade JR3 were initially identified from a grassland (Horz et al. 2005). These methanotrophs dominate in desert soils and are capable of oxidizing atmospheric CH_4 (Angel and Conrad 2009). *Methylocystis* are known to use CH_4 at a relatively low concentrations (Baani and Liesack 2008).

The first step of CH_4 oxidation is catalyzed by the enzyme methane monooxygenase (MMO) (Fig. 7.2) (Lee et al. 2013). This enzyme occurs as a membrane-bound particulate methane monooxygenase (pMMO) and (2) a cytoplasmic soluble methane monooxygenase (sMMO) (Ho et al. 2013). Most of the methanotrophs (except *Methylocella*) possess pMMO. This enzyme is constituted of three membrane-based polypeptides encoded by *pmoC*, *pmoA*, and *pmoB* (Kang and Lee 2016). Certain Type II methanotrophs (*Methylosinus*, *Methylocystis*), Type I methanotrophs (*Methylomonas*, *Methylomicrobium*), and Type X (*Methylococcus capsulatus*) possess sMMO in addition to pMMO (Cantera et al. 2016). The enzyme from the *Methylosinus*, *Methylocystis*, and *Methylococcus* has been thoroughly studied. The nucleotide sequence of the sMMO gene is constituted of *mmoX*, *mmoY*, *mmoB*, *mmoZ*, *mmoC*, and *mmoD* (Strand et al. 2013). The DNA sequences of this cluster are highly conserved. The *pmoA* gene encodes a 26-kDa subunit that harbors the active site for pMMO. The *mmoX* gene encodes for α-subunit of the sMMO hydroxylase. These genes are used as genetic markers to identify enzymes of various methanotrophs. Methanol dehydrogenase (MDH) is the second enzyme involved in methane oxidation. It is present in all methylotrophs including methane and methanol users. This enzyme is encoded by *mxaF* gene and is an appropriate marker for identifying methanotrophs possessing MDH activity (Haque et al. 2016).

7.2.2 Agricultural Management to Improve Atmospheric Methane Oxidation

Soil is the major biological sink for the atmospheric CH_4. The uncultivated methanotrophs are mostly responsible for atmospheric CH_4 (~1.8 ppm) consumption which is referred as "high-affinity" methane oxidation (HAMO). The HAMO is carried out by the conventional methanotrophs commonly found in paddy soil (Cai et al. 2016). In a study (Cai et al. 2016), it was observed that HAMO activity was quickly induced after low-affinity CH_4 oxidation. Low-affinity CH_4 oxidation was carried out at high CH_4 concentration (10,000 ppm). The high-affinity methanotrophic activity was lost over 2 weeks. However, the HAMO activity was regained by flush feeding the soil with 10,000 ppm of CH_4. The induction of HAMO activity occurred only after the rapid growth of methanotrophs. Metatranscriptome analysis revealed a strong transcriptional activity of the key enzymes of the conventional methanotrophs. This study highlighted that various intracellular polymers contribute toward HAMO activity. The study demonstrated that conventional methanotrophs are responsible for atmospheric CH_4 oxidation if the soil undergoes alternate draining.

7.2.3 Fe Redox Cycling can Modulate CH_4 Consumption in Soil

Our understanding on the relation between iron (Fe) reduction-oxidation (IRO) and CH_4 oxidation (consumption) is important to mitigate atmospheric CH_4. In a study (Mohanty et al. 2016), two soil types (alluvial and vertisol) were simulated to undergo microbial Fe reduction and aerobic oxidation repeatedly by natural wetting-drying cycle. Potential iron reduction rate k (μM Fe^{2+} produced g^{-1} soil d^{-1}) increased from 1.26 to 2.16 in vertisol and 1.95 to 3.05 in alluvial soil. Potential iron oxidation in both soils increased with repeated flooding and drying. The iron reduction-oxidation significantly ($p < 0.05$) stimulated CH_4 oxidation rate. The high-affinity CH_4 oxidation rate (μg CH_4 consumed per g soil per day) increased from 0.03 to 0.19. Low-affinity CH_4 oxidation rate increased from 0.05 to 0.47 in vertisol. X-ray diffraction (XRD) revealed that diffraction intensity of Fe minerals (magnetite and goethite) decreased over iron reduction-oxidation cycle. Real-time PCR quantification of methanotrophs (*pmoA* gene) confirmed that iron reduction-oxidation cycle stimulated ($p < 0.05$) methanotroph abundance. The study thus highlights that iron reduction-oxidation cycles can significantly enhance CH_4 oxidation in tropical soils. Previously, in a nitrate dependent Fe reduction-oxidation cycling study, it was found that highly reactive, amorphous Fe^{3+} oxide phases were formed (Fortin and Langley 2005). Surface area of Fe minerals is the gross indicator of the relative abundance of Fe^{3+} oxide surface available for microbial attachment (Tobler et al. 2007). This apparent dependence on surface area provides a functional explanation for the major differences in the microbial activity on various types of Fe^{3+} oxides (Roden and Zachara 1996). A high surface area of Fe minerals may change the soil environment to more aerobic and nutrient rich supporting the microbial activity (Li et al. 2013). It is hypothesized that low crystalline Fe minerals act as microenvironments for bacterial activity. Probably, these altered properties of Fe minerals results after iron reduction-oxidation cycling favored methanotrophs and stimulated CH_4 oxidation.

7.2.4 Ammonia-Oxidizing Bacteria

The lithoautotrophic ammonia-oxidizing bacteria are important in terms of their role in greenhouse gas emission. These bacteria use ammonia as sole energy source and are able to fix CO_2 through the Calvin-Benson cycle (Bock and Koops 2000). The ammonia-oxidizing bacteria are relevant for CH_4 consumption. Two main genera of ammonia oxidizers are *Nitrosomonas* and *Nitrosospira*. Both the genera belong to the β-proteobacteria. *Nitrosococcus* cluster belong phylogenetically within the γ-proteobacteria (Koops et al. 2000). The first step of nitrification is the oxidation of ammonia to hydroxylamine. This process is catalyzed by the ammonium monooxygenase. This enzyme is evolutionarily related to the methane monooxygenase enzyme (MMO) (Holmes et al. 1994). Often, the ammonium monooxygenase doesn't show high substrate specificity and oxidizes several compounds such as carbon monoxide and hydrocarbons (Hooper et al. 1998). This enzyme is also capable of oxidizing CH_4; however, it occurs at much lower pace than the methane monooxygenase (Bedard and Knowles 1989; Bodelier and Frenzel 1999).

7.3 Nitrous Oxide (N_2O) Cycling in Soil

Among the three different GHGs, N_2O is the most potent one. In terrestrial ecosystem it is produced from both natural and anthropogenic sources. Many other sources are there which produce significant amount of N_2O, but they are not clearly understood and also difficult to measure. Therefore, there is a general agreement that the atmospheric sources and sinks of N_2O are difficult to balance. Nitrous oxide is a long-lived trace gas, with its average mixing ratio of 330 ppbv (Arevalo-Martinez et al. 2013). The concentration of atmospheric N_2O has increased by 19% since preindustrial period but has increased by 0.77 ppb yr^{-1} during 2000–2009. It is a potential GHG with a 100-year global warming potential of 298 times higher than CO_2. It contributes 6.24% to the overall global climate change (Huang et al. 2013). The dominant sources of N_2O are the microbial processes in soils, sediments, and water bodies. N_2O emission from agricultural use of N fertilizer and manure management accounts for 4.3–5.8 Tg N_2O–N yr^{-1}. Its emission from natural soils ranges from 6 to 7 Tg N_2O–N yr^{-1}. Thus N_2O represents 56–70% of all global N_2O sources (Reay et al. 2012).

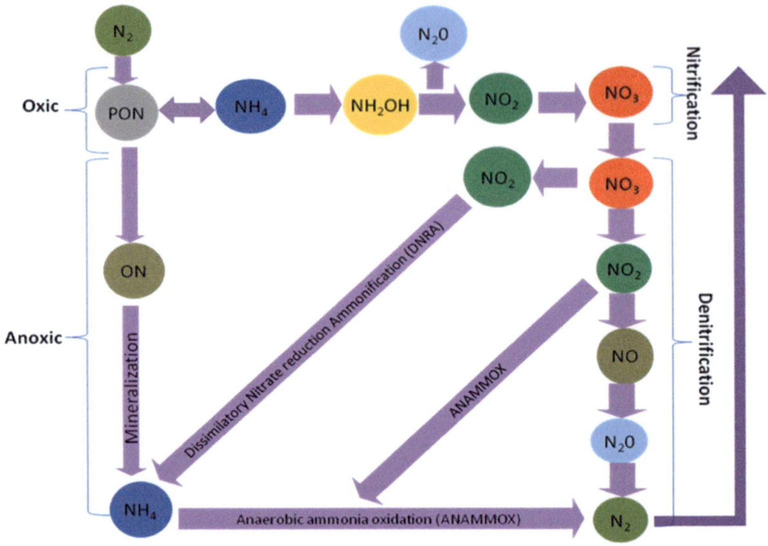

Fig. 7.3 Schematic illustration of N transformation (nitrification and denitrification) pathway and nitrous oxide production processes in soil

7.3.1 Nitrous Oxide Production from Nitrification and Denitrification

Estimation of N_2O budget at national and regional scales from local studies is a challenging task. This is mainly due to the variability of N_2O production caused by a multitude of interacting factors. Soil N_2O emission varies spatiotemporally and is also characterized by hot spots and timings. N_2O fluxes from soil are not only restricted to the specific sites of N fertilization but also owed to the volatilization, leaching, atmospheric deposition, and erosion processes. In natural ecosystem N fertilization creates new hot spots for N_2O emissions. However, it is challenging to integrate N_2O flux originating from nitrification and/or denitrification which often occurs in close vicinity. A substantial part of the NO_3^- formed by nitrification diffuse into anaerobic zone where it is denitrified into N_2, N_2O production occurs from both nitrification and denitrification (Fig. 7.3). N_2O from nitrification and denitrification contributes approximately 70% to the global N_2O budget (Butterbach-Bahl et al. 2013). The microbial N cycle with various metabolic steps is illustrated in Table 7.1.

The mechanism of N_2O production by nitrification is not clearly known. Three main hypotheses have been proposed:

1. During nitrification a constant proportion of NH_4^+ is converted to N_2O. This results into formation of various intermediate products. N_2O is produced from an intermediate product HNO produced during the oxidation of NH_2OH to NO_2^-.

Table 7.1 Various nitrogen-transforming microbial processes under oxic and anoxic ecosystem. N_2O is produced by both aerobic nitrifiers and anaerobic denitrifiers

		N molecules formed during microbial metabolism										
N cycling processes	Oxic/anoxic	$R\text{-}NH_2$	NH_4	NH_2OH	N_2O	NO	NO_2	NO_3	NO_2	NO	N_2O	N_2
Heterotrophic nitrification	Oxic	x	x	x	x	X	x	x				
Autotrophic NH_4 oxidation	Oxic		x	x	x	X	x	x				
Autotrophic NO_2 oxidation	Oxic						x	x				
Coupled nitrification-denitrification	Anoxic							x	x	X	x	X
Nitrifier denitrification	Anoxic						x			X	x	X
Anoxic denitrification	Anoxic							x	x	X	x	X
Denitrification with NO	Anoxic	x								X	x	X
Denitrification with N_2O	Anoxic	x								X	x	X
Dissimilatory NO_3 reduction ammonification (DNRA)	Anoxic		x				x	x			x	X
Biological N fixation	Anoxic	x									x	X
Anaerobic ammonia oxidation (ANAMMOX)	Anoxic		x				x					X

HNO is further oxidized to an unknown compound, which is subsequently oxidized to NO_2^-.
2. N_2O is produced when NO_2^- is reduced by accepting electron. Mostly, it occurs during NH_4^+ oxidation when O_2 pressure is low. Partial pressure of O_2 in soil varies with soil moisture.
3. N_2O is also produced during the partial oxidation of NH_4^+ into NO_2^-. When NO_2 is diffused into anaerobic regions of soil, it is denitrified to N_2O.

Both ammonia oxidizers and methanotrophs produce N_2O during the oxidation of NH_2OH to NO_2^-. Certain ammonia oxidizers reduce NO_2^- to N_2O and then to N_2 under anoxic condition. This process is termed as nitrifier denitrification. Under strict anaerobiosis N_2O is also produced by denitrifying organisms.

N_2O emission from soil is influenced by agricultural practices, climatic conditions, and soil properties. Soil factors include soil moisture and temperature, aeration, ammonium, and nitrate concentration, and pH. Soil moisture content is one of the predominant factors regulating N_2O emission from soils. However, it has been observed that alteration in the soil water content due to wetting events such as irrigation and rainfall can stimulate nitrification and denitrification and promote N_2O production. N_2O emission is highly correlated with water-filled pore space (WFPS). In an intensively managed calcareous fluvo-aquic soil, the highest N_2O emission occurred under 70% WFPS originating from both nitrification (35–53%) and denitrification (44–58%) (Huang et al. 2014). The favorable conditions for N_2O production from nitrification occur within the range of 30–70% WFPS (Hu et al. 2015), whereas denitrification dominates N_2O production in wet soils with >80–90% WFPS (Huang et al. 2014).

N_2O and other N compounds released from various N cycling microbial processes are given in Table 7.1. Two groups of nitrifiers including ammonia-oxidizing archaea (AOA) and ammonia-oxidizing bacteria (AOB) are responsible for the first step of nitrification (NH_3 to NO_2), and both groups constitute *amo*A gene encoding the alpha subunit of ammonia monooxygenase (AMO). However, the bacterial and archaeal *amo*A genes vary by their DNA sequences. The conversion of NO_2 to NO_3 is carried out by nitrite oxidoreductase encoded by *nxrB* gene. Until recently, AOB were believed to be the only nitrifier who carry out nitrification. However, AOA is being reported to have *amo*A genes. Studies reveal that AOA outnumber AOB in terrestrial environments.

Conversely, nitrification is a strictly aerobic process since the NH_4^+-oxidizing enzymes of different nitrifiers require O_2 (Weber et al. 2015). The effect of O_2 on nitrification and N_2O production has been studied well in pure microbial cultures. N_2O emission from nitrification takes place when the O_2 partial pressure is within the range of 0.1–0.5 kPa. Chemoautotrophic bacteria and archaea carry out nitrification, whereas denitrification is by heterotrophic bacteria (Table 7.1). Recently it has been found that many fungal strains also contribute to the N transformation. In an experiment, the role of fungi in NO_3 and N_2O production across the Southwestern United States was assessed. Soils were collected from desert and well-managed grassland sites. Differential role of bacteria and fungi on nitrification and N_2O

production was estimated by using selective inhibitors. Results highlighted that bacteria are responsible for nitrification and N_2O production in well-managed soils, while fungi are responsible for N_2O production in desert and semiarid soils.

Approximately two-thirds of total global N_2O emission comes from agricultural field soil. The intrinsic soil properties that control emission of N_2O are soil physical texture, pH, organic matter content, and available nitrogen. N_2O production in soil is generally attributed to microbiological processes, and all factors that modulate the activity of N_2O-producing microorganisms also regulate N_2O production. Recently, iron has been recognized as a regulator of N_2O emission from soil. Iron is known to involve in enzymatic and nonenzymatic reactions that mediate genesis of N_2O. The interrelation between iron and N_2O has not been clearly understood, and assumption of such relation has been neglected probably because iron is not routinely evaluated in soil samples in general agronomical research. Iron does not have a direct and immediate effect on the growth of crops or soil function. Also in soil iron does not significantly affect the immediate microbiological activity generally associated with N cycle and N_2O production. From series of experiments, it has been found that Fe content of soil is related to N_2O production. In soil at 50 and 100% moisture and NH_4-N fertilization, the significance of iron is high toward N_2O production (Huang et al. 2009). When ammonia is oxidized to hydroxylamine, the product reacts with Fe^{3+} to produce N_2O. Increased soil moisture stimulates mobility of solutes and enhances reaction between hydroxylamine and Fe^{3+} results into high N_2O production.

7.3.2 Mitigation of N_2O Emission from Soil

Nitrogen is the source of N_2O emission from agricultural soils. Minimization of N conversion to N_2O is the primary way to mitigate N_2O production from agricultural soil. There are few strategies to minimize N_2O emission. A study suggests using N-fixing bacteria to minimize N_2O emission (Itakura et al. 2013). Soybean is an important leguminous crop and hosts symbiotic nitrogen-fixing *Rhizobium* sp. In soybean fields, N_2O is also emitted but to a lesser extent. Emission of N_2O takes place during the degradation of the root nodules. Organic-N present in the nodules is mineralized to NH_4 which undergoes nitrification and denitrification and produce N_2O. This is carried out by expression of the *nosZ* gene. In a pure culture and vermiculite pot experiments, it was revealed that rhizobia strains with modified N_2O reductase gene (*nosZ*+ and *nosZ*++) lower N_2O emission. Therefore, these strains have been suggested as microbial inoculants for N_2O mitigation (Itakura et al. 2013).

Soil biochar amendment has been described as a promising tool to mitigate N_2O emission from agricultural soil. Many studies link the N_2O emission mitigation and the abundance and activity of N_2O-reducing microorganisms in biochar-amended soils. Biochar amendment shapes the N_2O reductase gene (*nosZ*) carrying soil microbial community. In a study, the diversity of bacterial 16S rRNA gene and *nosZ* genes was explored under the influence of biochar. Soil with biochar significantly altered the 16S rRNA gene-based community composition. Biochar amendment developed distinct bacterial community capable of N_2O reduction containing *nosZ* gene. The sequences of the enriched bacterial population were closely related to *nosZ* genes of *Pseudomonas stutzeri* and *Pedobacter saltans*. Further studies are needed to establish the molecular basis of nosZ gene expression in soil amended with biochar.

7.4 Conclusion

Current agricultural practice is likely to emit higher levels of GHGs to the atmosphere leading to increased global warming. Climate change eventually will have negative impact on the agricultural productivity. Soil microbes are the source and sink for atmospheric GHG. To regulate GHG emission from agriculture, it is important to understand the microbial processes governing the flux and feedback of GHGs. CH_4 and N_2O are the two most important GHGs emitted from agricultural soil. Several microbial processes are involved in GHG cycling like methanogenesis, methane oxidation, nitrification, and denitrification. CH_4 can be mitigated by promoting methanotrophs through soil biogeochemical process. One such intervention is alternate flooding and drying and Fe cycling. N_2O emission can be mitigated by exploring the microbial groups like N fixers with N_2O reductase gene. It is concluded that apart from agricultural management strategies, GHG emission from agriculture can be mitigated through biogeochemical processes and by using microbial groups with GHG-metabolizing genes.

Acknowledgment This manuscript is part of the project "Greenhouse gas emission from composting systems and characterization of GHG regulating microbes." Authors declare no conflict of interest of any type.

References

Angel R, Conrad R (2009) In situ measurement of methane fluxes and analysis of transcribed particulate methane monooxygenase in desert soils. Environ Microbiol 11:2598–2610

Angel R, Claus P, Conrad R (2012) Methanogenic archaea are globally ubiquitous in aerated soils and become active under wet anoxic conditions. ISME J 6:847–862

Arevalo-Martinez DL, Beyer M, Krumbholz M, Piller I, Kock A, Steinhoff T, Körtzinger A, Bange HW (2013) A new method for continuous measurements of oceanic and atmospheric N 2 O, CO and CO 2: performance of off-axis integrated cavity output spectroscopy (OA-ICOS) coupled to non-dispersive infrared detection (NDIR). Ocean Sci 9:1071–1087

Baani M, Liesack W (2008) Two isozymes of particulate methane monooxygenase with different methane oxidation kinetics are found in Methylocystis sp. strain SC2. Proc Natl Acad Sci 105:10203–10208

Ball BC (2013) Soil structure and greenhouse gas emissions: a synthesis of 20 years of experimentation. Eur J Soil Sci 64:357–373

Bao Q-L, Xiao K-Q, Chen Z, Yao H-Y, Zhu Y-G (2014) Methane production and methanogenic archaeal communities in two types of paddy soil amended with different amounts of rice straw. FEMS Microbiol Ecol 88:372–385

Bedard C, Knowles R (1989) Physiology, biochemistry, and specific inhibitors of CH4, NH4+, and CO oxidation by methanotrophs and nitrifiers. Microbiol Mol Biol Rev 53:68–84

Bock E, Koops HP (2000) The genus nitrobacter and related genera. Prokaryotes Dworkin M (ed) Springer, New York

Bodelier PLE, Frenzel P (1999) Contribution of Methanotrophic and nitrifying bacteria to CH4 and NH4+ oxidation in the rhizosphere of Rice plants as determined by new methods of discrimination. Appl Environ Microbiol 65:1826–1833

Butterbach-Bahl K, Baggs EM, Dannenmann M, Kiese R, Zechmeister-Boltenstern S (2013) Nitrous oxide emissions from soils: how well do we understand the processes and their controls? Phil Trans R Soc B 368:20130122

Cai Y, Zheng Y, Bodelier PL, Conrad R, Jia Z (2016) Conventional methanotrophs are responsible for atmospheric methane oxidation in paddy soils. Nat Commun 7:11728

Cantera S, Lebrero R, García-Encina PA, Muñoz R (2016) Evaluation of the influence of methane and copper concentration and methane mass transport on the community structure and biodegradation kinetics of methanotrophic cultures. J Environ Manag 171:11–20

Conrad R (1999) Contribution of hydrogen to methane production and control of hydrogen concentrations in methanogenic soils and sediments. FEMS Microb Ecol 28:193–202

Dong D, Yang M, Wang C, Wang H, Li Y, Luo J, Wu W (2013) Responses of methane emissions and rice yield to applications of biochar and straw in a paddy field. J Soils Sediments 13:1450–1460

Eller G, Deines P, Grey J, Richnow H-H, Krüger M (2005) Methane cycling in lake sediments and its influence on chironomid larval δ13C. FEMS Microbiol Ecol 54:339–350. https://doi.org/10.1016/j.femsec.2005.04.006

Forster P, Ramaswamy V, Artaxo P, Berntsen T, Betts R, Fahey DW, Haywood J, Lean J, Lowe DC, Myhre G (2007) Changes in atmospheric constituents and in radiative forcing. Clim Chang 20

Fortin D, Langley S (2005) Formation and occurrence of biogenic iron-rich minerals. Earth-Sci Rev 72:1–19

Han X, Sun X, Wang C, Wu M, Dong D, Zhong T, Thies JE, Wu W (2016) Mitigating methane emission from paddy soil with rice-straw biochar amendment under projected climate change. Sci Rep 6:1–10. https://doi.org/10.1038/srep24731

Haque MFU, Gu W, DiSpirito AA, Semrau JD (2016) Marker exchange mutagenesis of mxaF, encoding the large subunit of the Mxa methanol dehydrogenase, in Methylosinus trichosporium OB3b. Appl Environ Microbiol 82:1549–1555

Hernández M, Conrad R, Klose M, Ma K, Lu Y (2017) Structure and function of methanogenic microbial communities in soils from flooded rice and upland soybean fields from Sanjiang plain, NE China. Soil Biol Biochem 105:81–91

Ho A, Lüke C, Reim A, Frenzel P (2013) Selective stimulation in a natural community of methane oxidizing bacteria: effects of copper on pmoA transcription and activity. Soil Biol Biochem 65:211–216. https://doi.org/10.1016/j.soilbio.2013.05.027

Holmes RM, Fisher SG, Grimm NB (1994) Parafluvial nitrogen dynamics in a desert stream ecosystem. J N Amer Benthol Soc 13:468–478

Hooper K, Petreas MX, Chuvakova T, Kazbekova G, Druz N, Seminova G, Sharmanov T, Hayward D, She J, Visita P, Winkler J, McKinney M, Wade TJ, Grassman J, Stephens RD (1998) Analysis of breast milk to assess exposure to chlorinated contaminants in Kazakstan: high levels of 2,3,7, 8-tetrachlorodibenzo-p-dioxin (TCDD) in agricultural villages of southern Kazakstan. Environ Health Perspect 106:797–806

Horz HP, Rich V, Avrahami S, Bohannan BJM (2005) Methane-oxidizing bacteria in a California upland grassland soil: diversity and response to simulated global change. Appl Environ Microbiol 71:2642–2652

Hu H-W, Chen D, He J-Z (2015) Microbial regulation of terrestrial nitrous oxide formation: understanding the biological pathways for prediction of emission rates. FEMS Microbiol Rev 39:729–749

Huang B, Yu K, Gambrell RP (2009) Effects of ferric iron reduction and regeneration on nitrous oxide and methane emissions in a rice soil. Chemosphere 74:481–486

Huang T, Gao B, Christie P, Ju X (2013) Net global warming potential and greenhouse gas intensity in a double-cropping cereal rotation as affected by nitrogen and straw management. Biogeosciences 10:7897–7911

Huang T, Gao B, Hu X-K, Lu X, Well R, Christie P, Bakken LR, Ju X-T (2014) Ammonia-oxidation as an engine to generate nitrous oxide in an intensively managed calcareous Fluvo-aquic soil. Sci Rep 4

Itakura M, Uchida Y, Akiyama H, Hoshino YT, Shimomura Y, Morimoto S, Tago K, Wang Y, Hayakawa C, Uetake Y (2013) Mitigation of nitrous oxide emissions from soils by Bradyrhizobium japonicum inoculation. Nat Clim Chang 3:208–212

Kang TJ, Lee EY (2016) Metabolic versatility of microbial methane oxidation for biocatalytic methane conversion. J Ind Eng Chem 35:8–13

Knief C, Lipski A, Dunfield PF (2003) Diversity and activity of Methanotrophic bacteria in different upland soils. Appl Environ Microbiol 69:6703–6714. https://doi.org/10.1128/AEM.69.11.6703-6714.2003

Knief C, Kolb S, Bodelier PLE, Lipski A, Dunfield PF (2006) The active methanotrophic community in hydromorphic soils changes in response to changing methane concentration. Environ Microbiol 8:321–333

Knox SH, Sturtevant C, Matthes JH, Koteen L, Verfaillie J, Baldocchi D (2015) Agricultural peatland restoration: effects of land-use change on greenhouse gas (CO2 and CH4) fluxes in the Sacramento-san Joaquin Delta. Glob Chang Biol 21:750–765

Koops HP, Purkhold U, Pommerening-Roser A, Timmermann G, Wagner M (2000) The lithoautotrophic ammonia-oxidizing bacteria. Prokaryotes Dworkin M (ed). New York, Springer

Lee SJ, McCormick MS, Lippard SJ, Cho U-S (2013) Control of substrate access to the active site in methane monooxygenase. Nature 494:380–384

Li Y-L, Konhauser KO, Kappler A, Hao X-L (2013) Experimental low-grade alteration of biogenic magnetite indicates microbial involvement in generation of banded iron formations. Earth Planet Sci Lett 361:229–237

Lohila A, Aalto T, Aurela M, Hatakka J, Tuovinen J-P, Kilkki J, Penttilä T, Vuorenmaa J, Hänninen P, Sutinen R (2016) Large contribution of boreal upland forest soils to a catchment-scale CH4 balance in a wet year. Geophys Res Lett 43:2946–2953

Mau S, Blees J, Helmke E, Niemann H, Damm E (2013) Vertical distribution of methane oxidation and methanotrophic response to elevated methane concentrations in stratified waters of the Arctic fjord Storfjorden (Svalbard, Norway). Biogeosciences 10:6267–6278

Mohanty SR, Bandeepa GS, Dubey G, Ahirwar U, Patra AK, Kollah B (2016) Methane oxidation in response to iron reduction-oxidation metabolism in tropical soils. Eur J Soil Biol 78:75–81

Raupach MR, Marland G, Ciais P, Le Quéré C, Canadell JG, Klepper G, Field CB (2007) Global and regional drivers of accelerating CO2 emissions. Proc Natl Acad Sci 104:10288

Reay DS, Davidson EA, Smith KA, Smith P, Melillo JM, Dentener F, Crutzen PJ (2012) Global agriculture and nitrous oxide emissions. Nat Clim Chang 2:410–416

Rice AL, Butenhoff CL, Teama DG, Röger FH, Khalil MAK, Rasmussen RA (2016) Atmospheric methane isotopic record favors fossil sources flat in 1980s and 1990s with recent increase. Proc Natl Acad Sci 113:10791–10796

Ricke P, Erkel C, Kube M, Reinhardt R, Liesack W (2004) Comparative analysis of the conventional and novel pmo (particulate methane monooxygenase) operons from Methylocystis strain SC2. Appl Environ Microbiol 70:3055–3063. https://doi.org/10.1128/AEM.70.5.3055-3063.2004

Rissanen AJ, Karvinen A, Nykänen H, Peura S, Tiirola M, Mäki A, Kankaala P (2016) Effects of inorganic electron acceptors on methanogenesis and methanotrophy and on the community structure of bacteria and archaea in sediments of a boreal lake, in: EGU General Assembly Conference Abstracts, p. 3726

Roden EE, Zachara JM (1996) Microbial reduction of crystalline iron(III) oxides: influence of oxide surface area and potential for cell growth. Environ Sci Technol 30:1618–1628

Steger K, Premke K, Gudasz C, Boschker HTS, Tranvik LJ (2015) Comparative study on bacterial carbon sources in lake sediments: the role of methanotrophy. Aquat Microb Ecol 76:39–47

Strand S, Bruce N, Rylott L, Zhang L (2013) Phytoremediation of atmospheric methane. DTIC Document

Tate KR (2015) Soil methane oxidation and land-use change–from process to mitigation. Soil Biol Biochem 80:260–272

Tobler NB, Hofstetter TB, Straub KL, Fontana D, Schwarzenbach RP (2007) Iron-mediated microbial oxidation and abiotic reduction of organic contaminants under anoxic conditions. Environ Sci Technol 41:7765–7772

Ťupek B, Minkkinen K, Pumpanen J, Vesala T, Nikinmaa E (2014) CH_4 and N_2O dynamics in the boreal forest–mire ecotone. Biogeosci Discuss 11(6):8049–8084

Vanitchung S, Chidthaisong A, Conrad R (2014) Methane uptakes and emissions in upland tropical Forest and agricultural soils. J Sustain Energy Environ 5

Weber EB, Lehtovirta-Morley LE, Prosser JI, Gubry-Rangin C (2015) Ammonia oxidation is not required for growth of Group 1.1 c soil Thaumarchaeota. FEMS Microbiol Ecol 91. fiv001

Yuan J, Ding W, Liu D, Xiang J, Lin Y (2014a) Methane production potential and methanogenic archaea community dynamics along the Spartina Alterniflora invasion chronosequence in a coastal salt marsh. Appl Microbiol Biotechnol 98:1817–1829

Yuan Q, Pump J, Conrad R (2014b) Straw application in paddy soil enhances methane production also from other carbon sources. Biogeosciences 11:237–246

Zhang B, Pang C, Qin J, Liu K, Xu H, Li H (2013) Rice straw incorporation in winter with fertilizer-N application improves soil fertility and reduces global warming potential from a double rice paddy field. Biol Fertil Soils 49:1039–1052

Microbe-Mediated Bioremediation: An Eco-friendly Sustainable Approach for Environmental Clean-Up

8

Seema Sangwan and Ajinath Dukare

Abstract

Bioremediation provides a technique for cleaning up pollution by enhancing the natural biodegradation processes. Due to escalation in the costs of physical and chemical treatments, Microbe-mediated eco-friendly bioremediation technologies are getting more attractive. Each approach of bioremediation process has certain specific advantages and disadvantages, which need to be considered for each location. Microbial cellular enzyme-mediated remediation for successful degradation and clean-up of the wide range of organic contaminants in the polluted ecosystem is also novel and efficient approach. Numerous environmental factors limit and affect the efficiency of microbial degradation of xenobiotic pollutants in contaminated sites. The biological response to environmental pollutants varies within a microbial guild, and the presence of co-contaminants can elicit variable responses to the process of bioremediation. Regardless of which aspect of bioremediation is followed, this technology offers an efficient and cost-effective way to treat contaminated soil and groundwater.

Keywords

Bioremediation · Microbes · Pollution · Environment · Soil

S. Sangwan (✉)
Division of Microbiology, ICAR-IARI, New Delhi, India
e-mail: seemasangwan@iari.res.in

A. Dukare
Division of Horticultural Crop Processing, ICAR-CIPHET, Abohar, Punjab, India

T. K. Adhya et al. (eds.), *Advances in Soil Microbiology: Recent Trends and Future Prospects*, Microorganisms for Sustainability 3,
https://doi.org/10.1007/978-981-10-6178-3_8

8.1 Introduction

The rapid growth in agriculture and industry has resulted in the production, release and accumulation of large amounts of toxic xenobiotic compounds in the earth's environment (soil, air and water) that has become a cause for global concern (Gianfreda and Rao 2008). Primarily, xenobiotics are those toxic chemical compounds that are alien to the biological system and have tendency to gather and biomagnify in the earth's biosphere. These xenobiotic chemicals some of whom might be recalcitrant are directly released into the biosphere through wastewater discharge and release of solid residue from the various industries like chemical and pharmaceutical, plastics, paper and pulp mills, textile mills, agricultural operations (fungicides, herbicides and insecticides) (Lalithakumari 2011), etc. Indirect sources of xenobiotic residues include pharmaceutical compounds (Heberer 2002), pesticide residues (Karanth 2000) and anti-inflammatory drugs (Oaks et al. 2004).

Chemically, principal organic xenobiotics include alkanes, polycyclic aromatic hydrocarbons (PAHs), antibiotics, synthetic azo dyes, paints, insecticides and pesticides, fuels, various types of solvents, surfactants, pollutants (dioxins and polychlorinated biphenyls) and polyaromatic, chlorinated and nitro-aromatic compounds (Sinha et al. 2009). Concentration of these xenobiotic pollutants in natural environments has increased dramatically. For example, concentrations of PAHs in soil are varying from 1 μg to 300 g kg^{-1} soil, depending on the sources of contamination like fossil fuel combustion, gasification and liquefaction of coal, wood treatment processes and incineration of wastes (Bamforth and Singleton 2005). Besides, group of heavy metals including cadmium, chromium, copper, lead, mercury, nickel and zinc also act as primary inorganic chemical contaminants in soil and other natural ecosystem. Due to highly thermodynamic stability, these compounds are relatively largely persistent in the environment and lead to bioaccumulation or biomagnification in food chain. Further, these toxic compounds also have harmful effects on human health due to their carcinogenic, mutagenic and teratogenic effects. Xenobiotic compounds like biphenyl compounds, phenols and phthalates work as endocrine disruptors (Nagao 1998; Borgeest et al. 2002), while organochlorine pesticides like lindane (γ-HCH) affect the nervous system, liver and kidneys. Thus, due to overall harmful damage posed by the xenobiotic compound, various strategies have been developed for decontamination, sequestration and removal of these toxic contaminants from the biospheres (Saleem et al. 2008).

Presently, various chemical, physical and biological methods have been employed for control or removal of the toxic xenobiotic pollutants in soils. Commonly used non-biological pollutant decontamination methods (e.g. land-filling, recycling, pyrolysis and incineration) also had adverse effects on the environment due to formation of toxic intermediates in the process (Debarati et al. 2005). Furthermore, these methods are more expensive and sometimes difficult to execute, especially in non-point pollution, for instance, pesticides (Jain et al. 2005). Due to all these reasons, biological methods based on microbial technology have received much attention for remediation and degradation of toxic xenobiotic compounds (Gianfreda and Rao 2008). In the bioremediation process, specific biological system is deployed for either removal or transformation of toxic contaminants into nonhazardous products. Microbe-based bioremediation approaches have several advantages over physicochemical remediation methods such as cost-effective, environment-friendly convenient, complete degradation of organic pollutants and no collateral destruction of the site material or its indigenous flora and fauna (Timmis and Pieper 1999). Exploitation of the ability of microorganisms to remove toxic pollutants from contaminated sites is the most promising alternative treatment strategy (Finley et al. 2010).

8.2 Bioremediation: Principles Involved

Bioremediation is broadly described as the use of living biological organism including microorganisms or their enzymes or plants (phytoremediation) to remove, detoxify or reduce the concentration of toxic xenobiotic pollutant in environment.

Bioremediation offers the possibility of degrading, altering, removing, immobilizing or detoxifying toxic chemicals present in the earth's biospheres through the action of various microorganisms such as bacteria (Chowdhury et al. 2008; Jha et al. 2011), fungus (Kenneth 1996; Cho et al. 2009; Kumar et al. 2009) and even higher plants. Microbial activity in the bioremediation process results in the cleaning of the environment, via complete degradation, mineralization, sequestration or removal of the toxic pollutants. Degradation means that the microorganisms decompose the large and chemically complex group of pollutants into small and simple harmless compounds with release of natural by-products such as carbon dioxide (CO_2), water (H_2O) or other nontoxic intermediate molecule (Fig. 8.1).

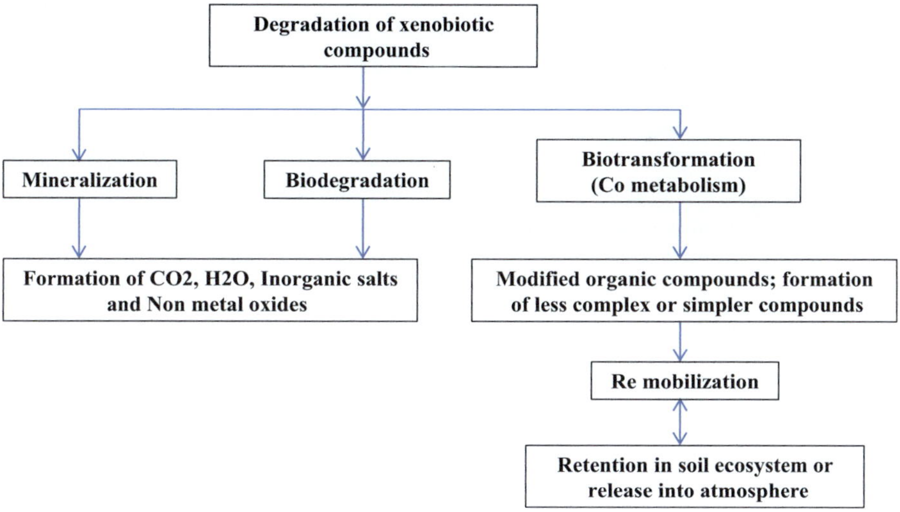

Fig. 8.1 Possible ways of biological transformations of xenobiotic compounds in the environment (Modified from Izabela (2002)

Sequestration is the mechanism by which toxic environmental contaminants are either trapped or altered in a way that makes it nontoxic or unavailable to living organisms. In the removal process, unsafe pollutant is not necessarily degraded, but the microbes physically remove it from the soil or water so that it can be collected and safely disposed. Generally, when compounds are persistent in the environment, bioremediation process proceeds through multiple steps of microbial enzyme systems or different microbial community and populations present at the contaminated sites. Agricultural soil, ground or surface waters, wastewater, aquatic bodies, sediments and air which are contaminated with release of toxic pollutants or chemicals are the sites where bioremediation process is employed (Ali Elredaisy 2010; Aghamiri et al. 2011).

Microbe-mediated degradation of xenobiotics is one of the important eco-friendly approaches to remove persistent harmful compounds accumulated in the environment. The ability of microorganisms to degrade, metabolize and transform xenobiotic compounds has been recognized as an efficient way of removing poisonous and detrimental wastes (Sridevi et al. 2011; Agarry and Solomon 2008). Microorganisms are ideally suited for the task of pollutant destruction and removal due to possession of enzyme system which allows them to use environmentally toxic pollutants as food and energy. Most of the advancements in bioremediation science has been attributed to the individual and interdisciplinary contribution provided by scientific areas of microbiology, molecular biology, biochemistry, analytical chemistry and environmental engineering (Sheehan 1997). Bioremediation process involves detoxification and mineralization, where the waste is converted into inorganic compounds such as carbon dioxide, water and methane (Reshma et al. 2011). Remediation of toxic pollutant from contaminated natural site through application of biological approaches has several advantages and disadvantages (Table 8.1).

8.3 Role of Microbes in Bioremediation

The ability of microbes to alter nearly all forms of organic material, their wider diversity and capabilities in catalytic mechanisms (Paul et al. 2005) and their ability to function even in the anaerobic and other extreme conditions (Mishra et al. 2001) makes them an attractive candidate for the bioremediation process. In addition, microbes play an important role in biogeochemical cycles and in sustainability of the biosphere ecosystems. The microbial transformation of xenobiotic pollutants can take place either in oxygenic or anoxygenic conditions. However, in the majority of cases, molecular oxygen participates in the first transformation reactions of both aliphatic and aromatic xenobiotic compounds (Cao et al. 2009; Sinha et al. 2009). Among the various group of microbes, bacteria have been found most efficient and dominant in the natural process of bioremediations. Under both aerobic and anaerobic conditions, bacteria have developed strategies for obtaining energy from virtually every compound by using electron acceptors such as ferric ions, nitrate, sulphate, etc. The diversity of microorganisms participating in the aerobic transformations of xenobiotic pollutants is vast (Table 8.2). Numerous bacterial

Table 8.1 Advantages and disadvantages of bioremediation (Sharma and Reddy 2004; Vivaldi 2001)

S. No.	Advantages	Disadvantages
1.	Offers possibility of complete breakdown/degradation/mineralization of organic pollutants into other nontoxic substances in natural ecosystem	May lead to the incomplete/partial degradation of organic contaminants resulting in the production of toxic intermediate compounds having more mobility than original contaminants
2.	Involves minimal requirement of external energy when compared to other remediation technologies/approaches	Being a naturally occurring microbial process, microbial activities in the contaminated sites are sensitive to the concentration of toxic compounds and various environmental factors (moisture, aeration, nutrient supply, temperature, pH, etc.)
3.	Depending upon conditions, can be implemented both as an in situ and ex situ method	Requires field monitoring to track the progress of biodegradation of the organic contaminants
4.	Relative simplicity of the technology, compared with many other on-site treatment technologies	Control of volatile organic compounds (VOCs) may become difficult if an ex situ bioremediation process is followed
5.	Cost of treatment per unit volume of soil or groundwater is very less as compared to other remediation methods	Process is lengthier and usually requires longer treatment time than other remediation technologies
6.	Complete degradation of xenobiotic pollutants is possible because the process does not involve pollutant transfer to other environmental medium/conditions	Not all xenobiotic compounds are susceptible to biodegradation; hence, the range of biodegradable contaminants that can be effectively treated is limited
7.	Minimal site disruption and low environmental impact and thus easily perceived positively by the public domain	Sometimes, residual levels of harmful intermediates can get too high (not meeting regulatory requirements), persistent and/or toxic in contaminated sites
8.	Requires low-technology equipment, i.e. readily available equipment	Due to lack of performance criteria regulations, performance evaluations are difficult because there is not a defined level of a "clean" site

genera, e.g. *Acinetobacter*, *Alcaligenes*, *Bacillus*, *Escherichia*, *Gordonia*, *Micrococcus*, *Moraxella*, *Pseudomonas*, *Pandoraea*, *Rhodococcus*, *Sphingobium* and *Streptomyces*, either individually or in combination are involved in the oxygenic breakdown, while bacterial genera involved in anaerobic degradation of xenobiotics include *Azoarcus*, *Clostridium*, *Desulfotomaculum*, *Desulfovibrio*, *Geobacter*, *Methanospirillum*, *Methanococcus*, *Methanosaeta*, *Pelotomaculum*, *Syntrophobacter*, *Syntrophus* and *Thauera* (Van Hamme et al. 2003; Kulkarni and Chaudhari 2007; Jindrova et al. 2002; Weelink et al. 2010).

Table 8.2 Representative examples of microorganisms involved in the bioremediation of xenobiotic compounds

Xenobiotic compound group	Name of xenobiotic pollutants	Degrading microorganisms	References
Polycyclic aromatic hydrocarbon compounds	Naphthalene	*Pseudomonas putida*	Habe and Omori (2003)
	Pyrene	*Mycobacterium* PYR-1	Kanaly and Harayama (2000)
	2,3,4-Chloroaniline	*Pseudomonas* sp.	Spain and Nishino (1987)
Pesticide compounds	Endosulphate compounds	*Arthrobacter* sp.	Weir et al. (2006)
	Endosulfan compounds	*Mycobacterium* sp.	Sutherland et al. (2002)
	DDT	*Dehalospirilum multivorans*	Chaudhry and Chapalamadugu (1991)
	2,4-D	*Alcaligenes eutrophus*	Don and Pemberton (1981)
Halogenated organic compounds	Vinyl chloride	*Dehalococcoides* sp.	He et al. (2003)
	PCE	*Dehalococcoides ethenogenes*195	Magnuson et al. (2000)
	Atrazine	*Pseudomonas* sp.	Bruhn et al. (1988)
Other xenobiotic compounds	PCB	*Rhodococcus* RHA1	Kimbara (2005)
	Dioxins		
	Benzene	*Dehalococcoides* sp.	Bunge et al. (2003)
		Dechloromonas sp.	Coates et al. (2001)
	Azo dyes	*Pseudomonas* sp.	Stolz (2001)
		Sphingomonas sp.	Reife and Freeman (2000)
		Xanthomonas sp.	
	Petroleum products	*Achromobacter* sp.	Austin et al. (1977)
		Micrococcus sp.	
		Bacillus sp.	
		Flavobacterium sp.	

8.4 Overview of Methods Employed for Remediation

US Environmental Protection Agency (US-EPA 2001, 2002), on the basis of removal and transportation of wastes for treatment, has described two methods of bioremediation, i.e. in situ and ex situ (Fig. 8.2). In situ bioremediation techniques involve the enhancement of indigenous microbial activity or inoculation of cultivated microbes into the contaminated environment, while ex situ treatment

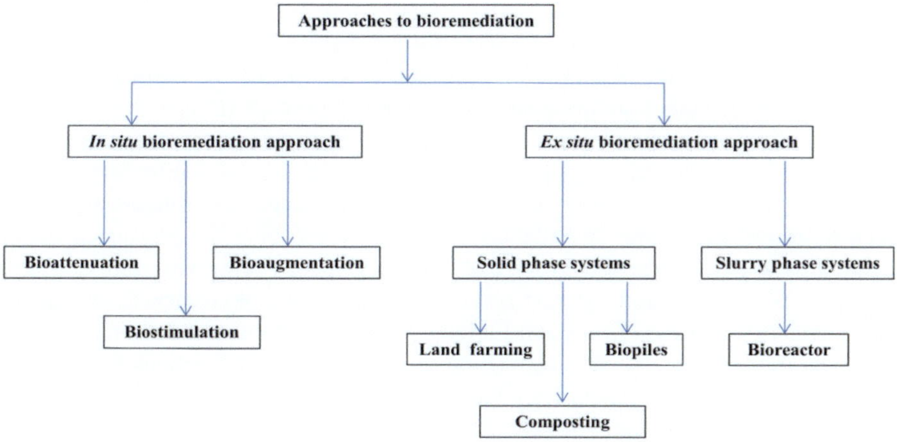

Fig. 8.2 An overview of methods deployed in microbial mediated bioremediation processes

techniques require removing contaminated soils or groundwater and treating them in a bioreactor or via surface treatment. While both in situ and ex situ remediation methods depend essentially on microbial metabolism, the in situ bioremediation methods are preferred to those of ex situ for ecological restoration of contaminated soil and groundwater environments (Jorgensen 2007).The appropriateness of a particular bioremediation technology is influenced by several factors, such as conditions of site, native population of microorganism and the type, quantity and toxicity of xenobiotic pollutant present.

8.4.1 In Situ Bioremediation Approach

This bioremediation approach does not require excavation or removal of contaminated soils or water in order to accomplish the process of remediation. In situ biodegradation generally involves supplying oxygen, nutrients and source of electron acceptors by circulating aqueous solutions through polluted soils to stimulate naturally occurring microorganisms' activity to degrade toxic organic pollutants. Microbial inoculum and cell-free enzymes have been used for in situ bioremediation. This approach has been mostly used for degradation of contaminants in saturated soils and groundwater (Vidali 2001; Evans and Furlong 2003). It is a superior method of cleaning contaminated environments since it is cheaper, uses native harmless microbial organisms to degrade the pollutants, is a safer method of degrading harmful compounds and is also possible to treat a large volume of contaminated soil or water with less release of toxic contaminants. Three different types of in situ microbial remediation methods are (i) bioattenuation, which is the natural degradation process; (ii) biostimulation, where premeditated stimulation of degradation of xenobiotic compounds is achieved by addition of water, nutrient, electron donors or acceptors; and (iii) bioaugmentation, where the microbial inoculum with demonstrated capabilities of degrading or transforming the chemical pollutants is added to the polluted environment (Madsen 1991).

In situ bioremediation can also be grouped as intrinsic bioremediation and engineered bioremediation. Intrinsic bioremediation approach deals with stimulation of indigenous or naturally occurring microbial activity by feeding them with stimulants (nutrients and oxygen), whereas engineered bioremediation involves the introduction of certain microorganisms to the contaminated site. Engineered bioremediation accelerates the degradation process by enhancing the physicochemical conditions to encourage the growth of microorganisms. Nutrients (nitrogen and phosphorus), oxygen and electron acceptors promote microbial growth and speed up remediation process (Evans and Furlong 2003). In situ bioremediation approach has been mostly used for the degradation of chlorinated hydrocarbons (Chiu et al. 2004; Ruppe et al. 2004), nitriles (Wang et al. 2004), nitrobenzenes (Coates et al. 2001; Rooney-Varga et al. 1999), anilines (Li et al. 2007), plasticizers (Cartwright et al. 2000) and other related compounds in soil and groundwater.

8.4.2 Ex Situ Bioremediation

Ex situ microbial bioremediation techniques require excavation of contaminated soil or pumping of polluted groundwater to facilitate process of degradation. This technique has more disadvantages than advantages. With the requirement of excavation of the contaminated samples for remediation treatment, the cost involved in ex situ bioremediation method can be high. In addition, the rate and consistency of the ex situ biodegradation process outcome can be different compared to in situ remediation methods. Depending on the state of the pollutant to be removed, ex situ bioremediation methods are classified as solid phase and slurry phase systems. The solid phase system involves treatments of various solids such as organic wastes, agricultural wastes, domestic wastes, industrial wastes, sewage sludge and municipal solid wastes. Solid phase treatment processes include land farming, composting and soil biopile techniques.

Land farming is a simple remediation method in which polluted soil is excavated and spread over a prepared bed and periodically tilled to stimulate indigenous biodegrading microorganisms for complete degradation or transformation of contaminants. It may be useful if only the upper 0.5 m of the soil profile is polluted. This bioremediation approach involves low monitoring and maintenance costs and, thus, has received much attention as a clean disposal option (US-EPA 2003). Composting is another surface treatment that has been successfully used to degrade, remove or transform toxic compounds. In composting approach, contaminated soil is combined with nonhazardous organic amendments such as manure or agricultural wastes to enhance the development of a rich microbial population and thus speed up the compost-mediated biodegradation process (Cunningham and Philip 2000). Biopiles are combination of land farming and composting bioremediation approaches. It is refined version of land farming aimed to minimize physical losses of the contaminants occurring due to leaching and volatilization. This method is mostly used for surface decontamination of petroleum hydrocarbon polluted sites. Biopiles provide a favourable environment for indigenous aerobic and anaerobic microflora involved in biodegradation (US-EPA 2003). Slurry phase is a relatively more rapid bioremediation process compared to the other treatment processes. This process is carried out in bioreactor/fermentor where contaminated soil is combined with water and indigenous microorganisms along with the addition of suitable nutrients and oxygen to control the optimum environment for the process of bioremediation. When the bioremediation treatment is completed, the liquid portion from solid is removed and disposed of properly (Cunningham and Philip 2000). Due to more manageable and predictive nature, rate and extent of biodegradation are greater in a bioreactor system than in situ or in solid phase systems.

8.5 Biochemical Pathways Involved in Bioremediation

Bacteria have the ability to readily incorporate the simple organic substances into their cells and metabolize them if favourable nutritional and environmental conditions are present. Some complex organic compounds which have longer molecular structures are degraded slowly, whereas some other compounds, also termed as recalcitrant or refractory compounds, are so complex that they cannot be degraded easily. There is a need for special techniques or integration of physico-chemical and biological techniques for effective remediation of these contaminants. Depending on the oxidation state of the pollutant, compounds can be either electron donors or electron acceptors in biodegradation process. The key players in bioremediation are the microbes as they generate the enzymes that catalyse the degradation reactions. Microbes use organic substances as a source of carbon and energy for their multiplication and maintenance and carry out various transformation reactions. The biodegradation pathways have been classified into three main categories on the basis of mechanisms used by microbes to gain energy, i.e. (1) aerobic, (2) anaerobic and (3) fermentation.

8.5.1 Aerobic Biodegradation

During aerobic degradation, organic compounds act as electron donor, and oxygen acts as electron acceptor. These oxygenation reactions also activate the substrate, and oxygen plays a crucial role in the aerobic degradation of aromatic compounds. The reaction has been described by Shimao (2001).

$$\text{Xenobiotic compound} + O_2 \rightarrow CO_2 + H_2O + \text{biomass} + \text{residue(s)}$$

Many aromatic and xenobiotic compounds like petroleum hydrocarbons, benzene, toluene, phenol, chlorinated aliphatics, chloroanilines, pentachlorophenol, dichlorobenzenes, naphthalene, fluorine, pyrene, etc. are rapidly and potentially degraded by microbes using aerobic degradation process. By producing enzymes, many bacterial consortia degrade toxic compounds to nontoxic forms by using them as substrates. The process of mineralization, i.e. the conversion of biodegradable compounds into simplest material, e.g. CO_2 and nitrogen compounds, is continued till all the biodegradable material is consumed by microbes and converted into CO_2 (Kyrikou and Briassoulis 2007) or biomass. Alkanes having straight structure with long carbon chains are easily degraded by aerobic pathways. In the aerobic degradation of alkanes, the terminal methyl group is oxidized into carboxylic acid, and then complete mineralization occurs through β-oxidation (Vander et al. 1992; Zhang and Bennett 2005). Molecular oxygen causes the oxygenation of aromatic compounds during aerobic degradation, and intermediates which are produced after this reaction directly enter into central metabolic pathways, e.g. Krebs cycle and β-oxidation (Wilson and Bouwer 1997; Sims and Overcash 1983). Subsequent fission of benzene ring occurs when microbes use oxygen to hydroxylate it during aerobic respiration. Enzymes, i.e. mono- and dioxygenase, which are involved in these processes incorporate one or two atoms of oxygen, respectively, into the ring. Dioxygenases are mainly involved in cleavages of the aromatic double bond located between two hydroxylated carbon atoms (ortho-pathway), adjacent to a hydroxylated carbon atom (meta-pathway) and indole ring (Hayaishi and Nozaki 1969). Aerobic biodegradation of benzene produces two intermediates, e.g. catechol, protocatechuate and gentisic acid, which are broken by simpler mechanisms as of simple acids and aldehydes and in turn are used as energy for cell synthesis and maintenance (Alexander 1977). Eaton (2001) and Hara et al. (2007) have reported the important role played by *Rhodococcus* RHA1 and *Arthrobacter keyseri* 12B bacteria in the degradation of 3,4-dihydroxybenzoate.

8.5.2 Anaerobic Biodegradation

In the absence of oxygen, anaerobic biodegradation takes place in which methane gas is produced instead of CO_2 (Swift 1998; Grima et al. 2002; Kyrikou and Briassoulis 2007). The anaerobic biodegradation reaction has been explained by Jayasekara et al. (2005) as follows:

$$\text{Xenobiotic compound} \rightarrow CO_2 + CH_4 + H_2O + \text{biomass} + \text{residue(s)}$$

Some of the pollutants are highly recalcitrant due to increase in halogenations so that they are not mineralized by aerobic degradation process. The electrophilicity of the molecule is increased due to the substitution of halogen, nitro and sulfo groups on the aromatic ring with which they resist the attack by oxygenases during aerobic degradation. Some of the examples are polychlorinated biphenyls (PCBs), chlorinated dioxins and some pesticides like DDT. Here reductive attacks by anaerobic bacteria have a high value and are of crucial importance to overcome the high persistence of halogenated xenobiotics from the biosphere. Reductive halogenations is performed by anaerobic bacteria either through gratuitous reaction or a new type of anaerobic respiration in which the degree of chlorination is reduced making the product more accessible for mineralization by aerobic bacteria (Van Agteren et al. 1998; Fritsche and Hofrichter 2008). During anaerobic biodegradation of PCBs (polychlorinated biphenyls), reductive dehalogenation is the first step where organic substrates act as electron donors. Electrons are accepted by PCBs which allow the anaerobic bacteria to transfer electrons to these compounds. Various xenobiotics which are present in different anaerobic habitats like water-laden soils, reticuloruminal contents, inter alia sediments, gastrointestinal contents, sludge digesters, feed-lot wastes, groundwater and landfill sites are degraded by anaerobic bacteria. Various anaerobic bacteria, e.g. *Bordetella*, *Acidovorax*, *Sphingomonas*, *Pseudomonas*, *Veillonella alcalescens*, *Variovorax*, *Geobacter metallireducens*, *Desulfuromonas michiganensis*, *Desulfovibrio* spp. and *Desulfitobacterium halogenans*, *D. oleovorans*, *D. acetonicum*, etc., have the capability to carry out anaerobic biodegradation. Anaerobic bacteria act on substituted and complex aromatic compounds which serve as electron acceptors accompanying with modifications of ring substituents (Gibson and Harwood 2002). The sulphate-reducing bacteria (SRB) which are obligate anaerobes represent a large group of anaerobic organisms which utilize sulphate as terminal electron acceptor during respiration and release hydrogen sulphide gas (Boetius et al. 2000; Sahrani et al. 2008), thus playing a crucial role in degradation of crude oil (Barton and Hamilton 2007).

8.5.3 Fermentation

There is one more mechanism through which microbes generate energy and carry out the biodegradation process. Some fermenting bacteria utilize the complex organic compounds as electron donors as well as electron acceptors and produce organic acids, alcohols, H_2 and CO_2, thus contributing to cleanliness of the environment.

8.6 Bioremediation Using Microbial Enzymes: A Cutting Edge Approach

Bioremediation is mainly dependent on the enzymes of degrading enzymes to carry out this process. In the new era of technology, use of enzymes as green, sustainable alternative for remediation of soil and groundwater is an emerging approach. Bioremediation using enzymes is a hybrid of biological and chemical treatment technology. Isolation and purification of these enzymes have become feasible due to the recent advancement in biology. These harvested enzymes are then injected into the contaminated water or soil. Enzyme remediation has been successfully evaluated overseas for clean-up of sites contaminated with wide range of organic contaminants, e.g. petroleum, PCBs and nitro phenols. Eliminating the long-term environmental liability, this enzymatic bioremediation does not generate any waste or need any disposal. There are a large number of enzymes produced by bacteria and fungi which have been reported to play an important role in the biodegradation of toxic organic pollutants. A cost-effective and nature-friendly biotechnology, bioremediation powered by microbial enzymes, is an advanced bioprocess technology.

Enzymes involved in bioremediation have been classified into various groups:

(i) ***Oxidoreductases:*** Oxidoreductases produced by many bacteria and fungi (Gianfreda et al. 1999) are the class of microbial enzymes which carry out oxidation-reduction reactions and detoxify various toxic organic compounds. Contaminants are converted into harmless compounds through these oxidative coupling reactions. Various microbial oxidoreductase enzymes have been exploited which cause the decoloration and degradation of azo dyes (Vidali 2001; Husain 2006) and detoxify various toxic xenobiotics such as phenolic or anilinic compounds through polymerization, copolymerization with other substrates or binding to humic substances (Park et al. 2006). Various phenolic compounds which are produced from decomposition of lignin in a soil environment are biodegraded by humification in which oxidoreductases play an important role. Many bacteria through their enzymes reduce the radioactive metals from an oxidized soluble form to a reduced insoluble form, thus minimizing their ill-effect (Leung 2004). Chlorinated phenolic compounds produced upon the partial degradation of lignin during pulp bleaching process are among the most ample recalcitrant wastes generated in the effluents of paper and pulp industry. Many fungal species produce extracellular oxidoreductases, e.g. laccase, manganese peroxidase and lignin.

Group of microbial oxidoreductases includes oxygenases which has the potential of metabolizing organic compounds by increasing their reactivity or water solubility or bringing about cleavage of the aromatic ring. With a broad range of substrates, these oxygenases are active against a wide range of compounds, including the chlorinated aliphatics and various halogenated organic compounds which comprise the largest groups of environmental pollutants including herbicides, insecticides, fungicides, hydraulic and heat transfer fluids, plasticizers and intermediates for chemical synthesis (Fetzner and Lingens 1994). These oxygenases are further divided into monooxygenases and dioxygenases.

Monooxygenases, which cause oxidation by incorporating one atom of oxygen molecule into their substrate, actively participate in dehalogenation, ammonification, desulfurization, denitrification and hydroxylation of various aromatic and aliphatic compounds (Arora et al. 2010). Among monooxygenases, methane monooxygenase is one of the best studied and well characterized which carry out biodegradation of hydrocarbon such as substituted methanes, alkanes, cycloalkanes, alkenes, haloalkenes, ethers, aromatic and heterocyclic hydrocarbons, etc. (Fox et al. 1990; Grosse et al. 1999). Dioxygenases catalyse enantiospecifically the oxygenation of wide range of aromatic compounds and therefore serve as part of nature's strategy for degrading them in the environment.

(ii) ***Laccases:*** A family of multicopper oxidases, laccases are produced by fungi, insects and bacteria. With concomitant reduction of molecular oxygen to water, these laccases catalyse the oxidation of a wide range of reduced phenolic and aromatic substrates (Gianfreda et al. 1999; Mai et al. 2000). Produced intra- and extracellularly, laccases are capable of catalysing the oxidation of ortho- and para-diphenols, aminophenols, polyphenols, polyamines, lignins and aryl diamines as well as some inorganic ions (Rodríguez Couto and Toca Herrera 2006). They also cause depolymerization of lignin, thus representing themselves as an ubiquitous group of enzymes which promise a great potential for biotechnological and bioremediation applications.

(iii) ***Peroxidases:*** Peroxidase is another group of ubiquitous enzymes which are produced by fungi and prokaryotes. Peroxidase removes chlorinated phenolic compounds from the contaminated environments (Rubilar et al. 2008). Peroxidases also catalyse the oxidation of lignin and other phenolic compounds at the expense of hydrogen peroxide (H_2O_2). Being haem or non-haem proteins, they play a crucial role in auxin metabolism, lignin and suberin formation, cross-linking of cell wall components, defence against pathogens or cell elongation in plants (Hiner et al. 2002; Koua et al. 2009). Depending upon the source and activity, peroxidases are divided into lignin peroxidase (LiP), manganese-dependent peroxidase (MnP) and versatile peroxidase (VP). All these have been widely studied due to their high potentiality to degrade toxic substances in nature.

(iv) ***Hydrolytic enzymes:*** Due to the extensive use of industrial chemicals and petroleum hydrocarbons, they have become a serious threat as hazardous contaminants in numerous aquatic and terrestrial ecosystems. Hydrolytic enzymes which disrupt major chemical bonds in toxic compounds offer an effective mechanism for the biodegradation of oil spill and organophosphate and carbamate insecticides. They play an important role in biodegradation of organochlorine insecticides such as DDT and heptachlor which are stable in well-aerated soil but readily degrade in anaerobic environments (Vasileva-Tonkova and Galabova 2003). The hemicellulase, cellulase and glycosidase have high potential usage due to their application in biomass degradation (Schmidt 2006).

(v) ***Lipases:*** Lipases are closely associated with the biodegradation of organic pollutants present in soil. They act on a variety of lipids and degrade them which are produced by microorganisms, animals and plants. Extracted from bacteria, actinomycetes and animal cell, lipases have been reported in the drastic reduction of total hydrocarbons from contaminated soil. Microbial lipases are more versatile because of their potent role in the bioremediation of oil spills (Margesin et al. 1999; Riffaldi et al. 2006), industrial wastes, triglycerides (Sharma et al. 2011) and hydrocarbon contaminants (Margesin et al. 1999; Riffaldi et al. 2006). By catalysing various reactions such as hydrolysis, interesterification, esterification, alcoholysis and aminolysis (Prasad and Manjunath 2011), they have been found to be most useful parameters for testing hydrocarbon degradation in the soil. Although they have diagnostic usage in bioremediation, its production cost has restricted its industrial use (Sharma et al. 2011; Joseph et al. 2006).

(vi) ***Cellulases:*** Most promising with potential of converting waste cellulosic material into foods to meet burgeoning population, microbial cellulases have been the subject of intense research (Bennet et al. 2002). These microbial cellulases can be cell bound, cell envelope associated and some extracellular cellulases. In a mixture of several enzymes, hydrolytic activity of cellulase enzyme is comprised of (a) endoglucanase which creates free chain ends by attacking cellulose fiber, (b) exoglucanase or cellobiohydrolase which degrades the cellulose molecule further by removing cellobiose units from the free chain ends and (c) β-glucosidase which hydrolyses cellobiose to glucose units. Cellulose degrades cellulose into reducing sugars which are further fermented by yeasts or bacteria into alcohol (Sun and Cheng 2002). Cellulases have also been employed for the release of ink in paper and pulp industry during recycling.

(vii) ***Proteases:*** Much proteinaceous substance gets their entry in the atmosphere due to shedding and moulting of appendages, as by-products of some industries like poultry, fishery, leather and also after death of animals. These proteinaceous substances are hydrolysed by the proteases which have been divided into two main groups, i.e. endopeptidase and exopeptidase depending upon their locus of activity on the substrate. Proteases have a wide range of applications in food, pharmaceutical, leather and detergent industry, thus playing direct and indirect role in bioremediation (Singh 2003; Beena and Geevarghese 2010).

8.7 Conclusion

The implementation of microbial bioremediation technology to decontaminate polluted sites is still a developing and evolving scientific approach. The mechanisms driving microbial metabolic activity and the degradation pathways of specific toxic xenobiotic pollutants need to be further elucidated before successful and better controlled site-specific remediation treatments are developed. With recent advances in molecular biotechnology, genetically modified organisms could be developed having higher capability and improved biodegradative performance. Apart from this, application of immobilized cells and enzymes represents new approaches that may help in the treatment of toxic xenobiotic contaminated site. A multidisciplinary scientific approach involving scientists and engineers is required to provide new strategies for refinement of available bioremediation methods. With the cooperation of soil microbiologists, chemists and engineers, it should be possible to decrease toxic pollutant concentrations at contaminated sites safely, efficiently and economically.

References

Agarry SE, Solomon BO (2008) Kinetics of batch microbial degradation of phenols by indigenous *Pseudomonas fluorescence*. Int J Environ Sci Technol 5(2):223–232

Aghamiri SF, Kabiri K, Emtiazi G (2011) A novel approach for optimization of crude oil bioremediation in soil by the Taguchi method. J Pet Environ Biotechnol 2:108

Alexander M (1977) Introduction to soil microbiology. Wiley, New York, pp 50–150

Ali Elredaisy SM (2010) Ecological benefits of bioremediation of oil contaminated water in rich Savannah of Palogue, upper nile area-Southern Sudan. J Bioremed Biodegrad 1:103. https://doi.org/10.4172/2155-6199.1000103

Arora PK, Srivastava A, Singh VP (2010) Application of monooxygenases in dehalogenation, desulphurization, denitrification and hydroxylation of aromatic compounds. J Bioremed Biodegrad 1:1–8

Austin B, Calomiris JJ, Walker JD, Colwell RR (1977) Numerical taxonomy and ecology of petroleum degrading bacteria. Appl Environ Microbiol 34:60–68

Bamforth SM, Singleton I (2005) Bioremediation of polycyclic aromatic hydrocarbons: current knowledge and future directions. J Chem Technol Biotechnol 80:723–736

Barton LL, Hamilton WA (2007) Sulphate reducing bacteria: environmental and engineered system. Cambridge University Press, Cambridge, p 558

Beena AK, Geevarghese PI (2010) A solvent tolerant thermostable protease from a psychrotrophic isolate obtained from pasteurized milk. Dev Microbiol Mol Biol 1:113–119

Bennet JW, Wunch KG, Faison BD (2002) Use of fungi for biodegradation. ASM Press, Washington, DC

Boetius A, Ravenschlag K, Schubert CJ, Rickert D, Widdel F, Gieseke A, Amann R, Jorgense BB, Witte U, Pfannkuche O (2000) A marine microbial consortium apparently mediating anaerobic oxidation of methane. Nature 407:623–626

Borgeest C, Greenfeld C, Tomic D, Flaws JA (2002) The effects of endocrine disrupting chemicals on the ovary. Front Biosci 7:1941–1948

Bruhn C, Batley RC, Knockmues HJ (1988) The in-vivo construction of 4-chloro-2- nitrophenol assimilatory bacteria. Arch Microbiol 150:171–177

Bunge M, Adrian L, Kraus A, Lorenz WG, Andreesen JR, Gorisch H, Lechner U (2003) Reductive dehalogenation of chlorinated dioxins by the anaerobic bacterium *Dehalococcoides ethenogenes* genes sp. strain CBDBI. Nature 421:357–360

Cao B, Nagarajan K, Loh KC (2009) Biodegradation of aromatic compounds: current status and opportunities for biomolecular approaches. Appl Microbial Biotechnol 85:207–212

Cartwright CD, Thompson IP, Burns RG (2000) Degradation and impact of phthalate plasticizers on soil microbial communities. Environ Toxicol Chem 19:1253–1261

Chaudhry GR, Chapalamadugu S (1991) Biodegradation of halogenated organic compounds. Microbiol Rev 55:59–79

Chiu TC, Yen JH, Liu TL, Wang YS (2004) Anaerobic degradation of the organochlorine pesticides DDT and heptachlor in river sediment of Taiwan. Bull Environ Contam Toxicol 72:821–828

Cho NS, Wilkolazka AJ, Staszczak M, Cho HY, Ohga S (2009) The role of laccase from white rot fungi to stress conditions. J Fac Agric Kyushu Univ 54:81–83

Chowdhury A, Pradhan S, Saha M, Sanyal N (2008) Impact of pesticides on soil microbiological parameters and possible bioremediation strategies. J Ind Microbiol 48:114–127

Coates JD, Chakraborty R, Lack JG, O'Connor SM, Cole KA, Bender KS, Achenbach LA (2001) Anaerobic benzene oxidation coupled to nitrate reduction in pure culture by two strains of *Dechloromonas*. Nature 411:1039–1043

Cunningham CJ, Philip JC (2000) Comparison of bioaugmentation and biostimulation in treatment of diesel contaminated soil, land contamination and reclamation. University of Edinburgh, Edinburgh

Debarati P, Gunjan P, Janmejay P, Rakesh VJK (2005) Accessing microbial diversity for bioremediation and environmental restoration. Trends Biotechnol 23:135–142

Don RH, Pemberton JM (1981) Properties of six pesticide degradation plasmids isolated from *Alcaligenes paradoxus* and *Alcaligenes eutrophus*. J Bacteriol 145:681–686
Eaton RW (2001) Plasmid-encoded phthalate catabolic pathway in *Arthrobacter keyseri* 12B. J Bacteriol 183:3689–3703
Evans GM, Furlong JC (2003) Environmental biotechnology theory and application. Wiley, West Sussex
Fetzner S, Lingens F (1994) Bacterial dehalogenases: biochemistry, genetics, and biotechnological applications. Microbiol Rev 58(4):641–685
Finley SD, Broadbelt LJ, Hatzimanikatis V (2010) In silico feasibility of novel biodegradation pathways for 1, 2, 4-trichlorobenzene. Bio Med Cent Syst Biol 4:1–14
Fox BG, Borneman JG, Wackett LP, Lipscomb JD (1990) Haloalkene oxidation by the soluble methane monooxygenase from *Methylosinus trichosporium* OB3b: mechanistic and environmental implications. Biochemistry 29(27):6419–6427
Fritsche W, Hofrichter M (2008) Aerobic, degradation, by microorganisms: principles of bacterial degradation. In: Rehm HJ, Reed G, Puhler A, Stadler A (eds) Biotechnology, environmental processes II, vol vol IIb. Wiley-VCH, Weinhein, pp 145–167
Gianfreda L, Rao MA (2008) Microbe-mediated bioremediation: an eco-friendly sustainable approach for environmental clean-up. Crit Rev Environ Sci Technol 38(4):269–310
Gianfreda L, Xu F, Bollag JM (1999) Laccases: a useful group of oxidoreductive enzymes. Bioremed J 3(1):1–25
Gibson J, Harwood CS (2002) Metabolic diversity in aromatic compound utilization by anaerobic microbes. Annu Rev Microbiol 56:345–369
Grima S, Bellon-Maurel V, Feuilloley P, Silvestre F (2002) Aerobic biodegradation of polymers in solid-state conditions: a review of environmental and physicochemical parameter settings in laboratory simulation. J Polym Environ 8(4):183–195
Grosse S, Laramee L, Wendlandt KD, McDonald IR, Miguez CB, Kleber HP (1999) Purification and characterization of the soluble methane monooxygenase of the type II methanotrophic bacterium *Methylocystis* sp. strain WI 14. Appl Environ Microbiol 65(9):3929–3935
Habe H, Omori T (2003) Genetics of polycyclic aromatic hydrocarbon degradation by diverse aerobic bacteria. Biosci Biotechnol Biochem 67:225–243
Hara H, Eltis LD, Davies JE, Mohn WW (2007) Transcriptomic analysis reveals a bifurcated terepthalate degradation pathway in *Rhodococcus* sp. strain RHA1. J Bacteriol 189:1641–1647
Hayaishi O, Nozaki M (1969) Nature and mechanism of oxygenases. Science 164(*3878*):389–396
He J, Ritalahti KM, Yang KL, Koenigsberg SS, Löffler FE (2003) Detoxification of vinyl chloride to ethane coupled to an anaerobic bacterium. Nature 424:62–65
Heberer T (2002) Occurrence, fate and removal of pharmaceutical residues in the aquatic environment: a review of recent research data. Toxicol Lett 131(1–2):5–17
Hiner NP, Ruiz JH, Rodri JN (2002) Reactions of the class II peroxidases, lignin peroxidase and *Arthromyces ramosus* peroxidase, with hydrogen peroxide: catalase-like activity, compound III formation, and enzyme inactivation. J Biol Chem 277(30):26879–26885
Husain Q (2006) Potential applications of the oxidoreductive enzymes in the decolorization and detoxification of textile and other synthetic dyes from polluted water: a review. Crit Rev Biotechnol 26(4):201–221
Izabela G (2002) Microbial transformation of xenobiotics. Chemik 66(8):835–842
Jain RK, Kapur M, Labana S, Lal L, Sarma PM, Bhattacharya D, Thakur IS (2005) Microbial diversity: application of microorganisms for the biodegradation of xenobiotics. Curr Sci 89 (1):101–112
Jayasekara R, Harding I, Bowater I, Lornergan G (2005) Biodegradability of selected range of polymers and polymer blends and standard methods for assessment of biodegradation. J Polym Environ 13:231–251
Jha AK, Singh K, Sharma C, Singh SK, Gupta PK (2011) Assessment of methane and nitrous oxide emissions from livestock in India. J Earth Sci Clim Chang 1:107

Jindrova E, Chocova M, Demnerova K, Brenner V (2002) Bacterial aerobic degradation of benzene, toluene, ethylbenzene and xylene. Folia Microbiol 47(2):83

Jorgensen KS (2007) In situ bioremediation. Adv Appl Microbiol 61:285–305

Joseph B, Ramteke PW, Kumar PA (2006) Studies on the enhanced production of extracellular lipase by *Staphylococcus epidermidis*. J Gen Appl Microbiol 52(6):315–320

Kanaly RA, Harayama S (2000) Biodegradation of high-molecular weight polycyclic aromatic hydrocarbons by bacteria. J Bacteriol 182:2059–2067

Karanth NGK (2000) Challenges of limiting pesticide residues in fresh vegetables: the Indian experience food safety management in developing countries. CIRAD-FAO, Montpellier, pp 11–13

Kenneth EH (1996) Extracellular free radical biochemistry of ligninolytic fungi. New J Chem 20:195–198

Kimbara K (2005) Recent developments in the study of microbial aerobic degradation of polychlorinated biphenyls. Microbes Environ 20:127–134

Koua D, Cerutti L, Falquet L (2009) PeroxiBase: a database with new tools for peroxidase family classification. Nucleic Acids Res 37(1):261–266

Kulkarni M, Chaudhari A (2007) Microbial remediation of nitro-aromatic compounds*: an review*. J Environ Manag 85:496

Kumar KK, Prasad MK, Sarma GVS, Murthy CVR (2009) Removal of Cd (II) from aqueous solution using immobilized *Rhizomucor Tauricus*. J Microbial Biochem Technol 1:015–021

Kyrikou J, Briassoulis D (2007) Biodegradation of agricultural plastic films: a critical review. J Polym Environ 15:125–150

Lalithakumari D (2011) Microbes: a tribute to clean environment. University of Madras, Chennai

Leung M (2004) Bioremediation: techniques for cleaning up a mess. J Biotechnol 2:18–22

Li W, Barrington S, Kim J (2007) Biodegradation of pentyl amine and aniline from petrochemical wastewater. J Environ Manag 83:191–197

Madsen EL (1991) Determining in situ biodegradation: facts and challenges. Environ Sci Technol 25:1663–1673

Magnuson JK, Romine MF, Burris DR, Kingsley MT (2000) Trichloroethene reductive dehalogenase from *Dehalococcoides ethenogenes*: sequence of *tce*A and substrate range characterization. Appl Environ Microbiol 66:5141–5147

Mai C, Schormann W, Milstein O, Huttermann A (2000) Enhanced stability of laccase in the presence of phenolic compounds. Appl Microbiol Biotechnol 54(4):510–514

Margesin R, Zimmerbauer A, Schinner F (1999) Soil lipase activity—a useful indicator of oil biodegradation. Biotechnol Tech 13(12):859–863

Mishra V, Lal R, Srinivasan S (2001) Enzymes and operons mediating xenobiotic degradation in bacteria. Crit Rev Microbiol 27:133–166

Nagao T (1998) Health effects of endocrine disruptors. Jpn J Toxicol Environ Health 44:151–167

Oaks JL, Gilbert M, Virani MZ, Watson RT, Meteyer CU, Rideout BA, Shivprasad HL, Ahmed S, Chaudhary MJ, Arshad M, Mahmood S, Ali A, Khan AA (2004) Diclofenac residues as the cause of vulture population decline in Pakistan. Nature 427:630–633

Park JW, Park BK, Kim JE (2006) Remediation of soil contaminated with 2,4-dichlorophenol by treatment of minced shepherd's purse roots. Arch Environ Contam Toxicol 50(2):191–195

Paul D, Pandey G, Pandey J, Jain RK (2005) Accessing microbial diversity for bioremediation and environmental restoration. Trends Biotechnol 23:135–142

Prasad MP, Manjunath K (2011) Comparative study on biodegradation of lipid-rich wastewater using lipase producing bacterial species. Indian J Biotechnol 10(1):121–124

Reife A, Freeman HS (2000) Pollution prevention in the production of dyes and pigments. Text Chem Color Am Dyes Rep 32:56–60

Reshma SV, Spandana S, Sowmya M (2011) Bioremediation technologies. World Congress of Biotechnology, India, 40

Riffaldi R, Levi-Minzi R, Cardelli R, Palumbo S, Saviozzi A (2006) Soil biological activities in monitoring the bioremediation of diesel oil-contaminated soil. Water Air Soil Pollut 170 (1–4):3–15

Rodríguez Couto S, Toca Herrera JL (2006) Industrial and biotechnological applications of laccases: a review. Biotechnol Adv 24(5):500–513

Rooney-Varga JN, Anderson RT, Fraga JL, Ringelberg D, Lovley DR (1999) Microbial communities associated with anaerobic benzene degradation in a petroleum-contaminated aquifer. Appl Environ Microbiol 65:3056–3063

Rubilar O, Diez MC, Gianfreda L (2008) Transformation of chlorinated phenolic compounds by white rot fungi. Crit Rev Environ Sci Technol 38(4):227–268

Ruppe S, Neumann A, Braekevelt E, Tomy GT, Stern GA, Maruya KA, Vetter W (2004) Anaerobic transformation of compounds of technical toxaphene. 2. Fate of compounds lacking geminal chlorine atoms. Environ Toxicol Chem 23:591–598

Sahrani FK, Ibrahim Z, Yahya A, Aziz M (2008) Isolation and identification of marine sulphate reducing bacteria, *Desolfovibrio* sp. and *Citrobacter freundii* from Pasir Gudang, Malaysia. Science 47:365–371

Saleem M, Brim H, Hussain S, Arshad M, Leigh MB, Zia UH (2008) Perspectives on microbial cell surface display in bioremediation. Biotechnol Adv 26:151–161

Schmidt O (2006) Wood and tree fungi. Springer, Berlin

Sharma HD, Reddy KR (2004) Geoenvironmental engineering. Wiley, Hoboken

Sharma D, Sharma B, Shukla AK (2011) Biotechnological approach of microbial lipase: a review. Biotechnology 10(1):23–40

Sheehan D (1997) Bioremediation protocols. Humana Press, Totowa

Shimao M (2001) Biodegradation of plastics. Curr Opin Biotechnol 12:242–247

Sims RC, Overcash MR (1983) Fate of polynuclear aromatic compounds (PNAs) in soil plant system. Residue Rev 88:1–68

Singh CJ (2003) Optimization of an extracellular protease of *Chrysosporium keratinophilum* and its potential in bioremediation of keratinic wastes. Mycopathologia 156(3):151–156

Sinha S, Chattopadhyay P, Pan I, Chatterjee S, Chanda P, Bandyopadhyay D, Das K, Sen SK (2009) Microbial transformation of xenobiotics for environmental bioremediation. Afr J Biotechnol 8(22):6016

Spain J, Nishino SF (1987) Degradation of 1, 4 dichlorobenzene by a *Pseudomonas* sp. Appl Environ Microbiol 53:1010–1019

Sridevi V, Lakshmi MV, Swamy AV, Rao MN (2011) Implementation of response surface methodology for phenol degradation using *Pseudomonas putida* (NCIM 2102). J Bioremed Biodegrad. https://doi.org/10.4172/2155-6199.1000121

Stolz A (2001) Basic and applied aspects in the microbial degradation of azo dyes. Appl Microbiol Biotechnol 56:69–80

Sun Y, Cheng J (2002) Hydrolysis of lignocellulosic materials for ethanol production: a review. Bioresour Technol 83(1):1–11

Sutherland TD, Horne I, Russell RJ, Oakeshott JG (2002) Isolation and characterization of a *Mycobacterium* strain that metabolizes the insecticide endosulfan. J Appl Microbiol 93:380–389

Swift G (1998) Requirements for biodegradable water soluble polymers. Polym Degrad Stab 59:19–24

Timmis KN, Pieper DH (1999) Bacteria designed for bioremediation. Trends Biotechnol 17:200–204

US-EPA (2001) Remediation case studies. Federal Remediation Technology Roundtable. Report No. 542- F-01-032

US-EPA (2002) Handbook on in situ treatment of hazardous waste contaminated soils

US-EPA (2003) Engineering issue: in situ and ex situ biodegradation technologies for remediation of contaminated sites:6–15

Van Agteren MH, Keuning S, Janssen DB (1998) Handbook on biodegradation and biological treatment of hazardous organic compounds. Kluwer, Dordrecht

Van Hamme JD, Singh A, Ward OP (2003) Recent advances in petroleum microbiology. Microbiol Mol Biol Rev 67(4):503

Vander JRM, Vos WMD, Harayama S, Zehnder AJB (1992) Molecular mechanisms of genetic adaptation to xenobiotic compounds. Microbiol Rev 56:677–694

Vasileva-Tonkova E, Galabova D (2003) Hydrolytic enzymes and surfactants of bacterial isolates from lubricant contaminated wastewater. Z Naturforsch 58(1–2):87–92

Vidali M (2001) Bioremediation: an overview. Pure Appl Chem 73(7):1163–1172

Vivaldi M (2001) Bioremediation: an overview. Padova, Italy

Wang CC, Lee CM, Chen LJ (2004) Removal of nitriles from synthetic wastewater by acrylonitrile utilizing bacteria. J Environ Sci Health Part A Tox Hazard Subst Environ Eng 39:1767–1779

Weelink SAB, van Eekert MHA, Stams AJM (2010) Degradation of BTEX by anaerobic bacteria: physiology and application. Rev Environ Sci Biotechnol 9:359

Weir KM, Sutherland TD, Horne I, Russell RJ, Oakeshott JG (2006) A single moonoxygenase, *ese*, is involved in the metabolism of the organochlorides endosulfan and endosulphate in an *Arthrobacter* sp. Appl Environ Microbiol 72:3524–3530

Wilson L, Bouwer E (1997) Biodegradation of aromatic compounds under mixed oxygen/ denitrifying conditions: a review. J Ind Microbiol Biotechnol 18:116–130

Zhang C, Bennett GN (2005) Biodegradation of xenobiotics by anaerobic bacteria. Appl Microbiol Biotechnol 67:600–618

Check for updates

Remediation of Oily Sludge- and Oil-Contaminated Soil from Petroleum Industry: Recent Developments and Future Prospects

9

Subhasis Das, Nanthakumar Kuppanan,
Veeranna A. Channashettar, and Banwari Lal

Abstract

Oily sludge is a significant solid waste generated from petroleum industry across the globe. These wastes are stored in designated pits near to the oil production facilities for longer period of time creating further contamination of bottom soil layers in the pits. These solid wastes are mainly recalcitrant residues characterized as a stable oil/water emulsion of water, soils, petroleum hydrocarbons, and metals. These wastes are considered as hazardous waste by many countries in the world which need an effective remediation technology for their treatment and disposal. This chapter focuses on the origin of oil exploration process and source of oily sludge, their characteristics and toxicity, recovery, and disposal of oily sludge. The use of bioremediation technology for disposal of oily sludge with case studies based on Kuwait is described.

Keywords

Remediation · PHC · Petroleum industry · Kuwait

9.1 Introduction

The long-term exploration, production, transportation, storage, refinery processes, and use of different petroleum derivatives create widespread contamination of soil generated by oily sludge in and around the facilities of petroleum industry. It contains higher concentrations of petroleum hydrocarbons and recalcitrant residues and classified as high priority pollutants. Many countries recognized the oily sludge

S. Das (✉) · N. Kuppanan · V. A. Channashettar · B. Lal
Environmental and Industrial Biotechnology Division, The Energy and Resources Institute, New Delhi, India
e-mail: subhasis.das@teri.res.in

T. K. Adhya et al. (eds.), *Advances in Soil Microbiology: Recent Trends and Future Prospects*, Microorganisms for Sustainability 3,
https://doi.org/10.1007/978-981-10-6178-3_9

as a hazardous waste and devised their own policies for disposal or treatment of these hazardous compounds. A successful remediation technology could be a global challenge due to its hazardous nature and sharp increase in quantity. During the last few decades, oily sludge was treated by several physical, chemical, and biological methods including thermal incineration, solidification, solvent extraction, ultrasonic treatment, photocatalysis, pyrolysis, and bioremediation. This chapter attempts to describe the sources of oily sludge and associated contaminated soil, characteristics of the sludge, best available remediation technologies, advantages and limitation of the technology, and global prospects and recent developments.

9.2 Origin, Characteristic, and Toxicity of Petroleum Wastes

9.2.1 Origin of Oil and Exploration Process

Biodegraded oils dominate the world petroleum reserves (Roadifer 1987) that are scattered across the globe predominantly in Middle East countries, foreland basins in America, Atlantic margin basins of Africa, South America, and Canada, and the Gulf of Mexico (Head et al. 2003). At temperatures up to about 80 °C, petroleum in subsurface reservoirs could be degraded biologically by microorganisms over the geological time scale. As a result, denser heavy oils are produced by breakdown of the hydrocarbons in subsurface reservoirs. Evidence of active microbial communities in petroleum reservoirs since the 1930s (Bastin 1926; Krejci-Graf 1932; Head et al. 2003) indicates that microorganisms inhabiting deep subsurface sediments receive nutrients and oxidants for their growth and proliferation from plate tectonic process. The petroleum industries produce and transport billion tons of heavy crude oil, natural gas, and their derivatives annually for the production of large quantities of refined product including liquid petroleum gases, kerosene, diesel, petrol, aviation fuel, gasoline, and lubricants. Few products are further used in other petroleum industry as feedstock for downstream products (Olajire 2014).

9.2.2 Sources of Oily Sludge and Contaminated Soil

Upstream operations including extraction, transportation, and storage of the heavy crude oil generate large amount of oily sludge. These oily wastes could be categorized either into waste oil or sludge based on water/oil in oil matrix (Hu et al. 2013). Complex mixture of hydrocarbons, water, metals, and suspended fine solids is the typical physical form of petroleum sludge (Elektorowicz and Habibi 2005). In upstream operations, slop oil at oil wells, crude oil storage tank bottom sediments, drilling mud residues, aboveground pipeline leakages, oil waste dumping in effluent and sludge pit, and oil spills are major sources of oily sludge (Olajire 2014). While in the downstream operation, slop oil emulsion solid, heat exchange bundle cleaning sludge, residues from oil/water separator of different types, sediments deposited at the bottom of transporting vehicles and storage tanks, sludge from flocculation-flotation unit, dissolved air flotation system, induced air flotation unit of refining system, and excess activated sludge from biological wastewater treatment plant are the major sources of oily sludge (Hu et al. 2013). The US-EPA reported that in the USA, each refinery generates 30,000 tons of oily sludge annually, and in China, the estimated production of oily sludge is three million tons annually (Hu et al. 2013). In India average annual volume of oily sludge generated by oil refineries is approximately 28,000 tons (Bhattacharyya and Shekdar 2003), whereas, in 2011–2012, oily sludge generated from Indian refineries are 47,000 tons (Kumar and Raj Mohan 2013). These oil-sediment-water mixture and neighboring contaminated soils need to be treated.

9.2.3 Oily Sludge/Soil Characteristics and Toxicity

The pH range of oily sludge is 6.5–7.5 depending upon its chemical compositions and sources. The other physical properties such as density, viscosity, and heat value could also change with its chemical composition. The PHC and other organic compounds found in crude oil and oily sludge are classified as SARA components composed of saturates of aliphatic and aromatic fractions, resins, and asphaltene. Aliphatic and aromatic fractions are predominant in any of the oily sludge contributing 75–85% followed by resins (more than 8–22%) and asphaltenes (8–10%). Major aliphatic and aromatic compounds are alkene, cycloalkene, benzene, toluene, xylene, naphthalene, phenol, and several polycyclic aromatic hydrocarbons (PAHs) including anthracene, benzofluorene, chrysene, phenanthrene, and pyrenes. The resins are mainly composed of nitrogen (3%), sulfur (0.3–10%), and oxygen (4.8%) contents. Some polar naphthenic acids, mercaptans, thiophenes, and pyridines are the major resin compounds. Apart from the above constituents, oily sludge also contains a variety of heavy metals such as zinc (Zn), lead (Pb), copper (Cu), nickel (Ni), and chromium (Cr).

The toxic substances present in the sludge pose serious threats to the environment. These substances directly enter into the soil and also contaminate the groundwater. Nutrient deficiency, improper seed germination, restricted growth, or demises of plant are the major drawbacks due to the presence of toxic substances. Reduced water retention is also a problem due to highly viscous nature of these substances which could clog the soil pores and covers the soil surfaces. Sometimes secondary metabolites after degradation of oily sludge could also be toxic and recalcitrant.

9.3 Overview of Sludge Remediation

Oily sludge could be remediated by three-tiered oily sludge waste management strategy (Hu et al. 2013) where technologies are used to reduce the quantity of oily sludge generation from operation in petroleum industry, recovery, and reuse of valuable fuels extracted from oily sludge. If both the technology failed then considered option is to dispose of not treatable fraction to the landfill sites (Pinheiro and Holanda 2009; da Silva et al. 2012; Hu et al. 2013).

9.3.1 Oil Recovery Technologies

Recycling and reuse of oily sludge is considered as a most desirable environmental and feasible option which has received increasing interest and enables petroleum industries to reuse the oil for recovery and reformulation of energy and nutrient recovery. Furthermore, recycling can reduce waste volume by 70% of hazardous waste and destruction of the pathogens and toxic organic compounds. As per API (API Environmental Guidance Document 1989), the hydrocarbon recovery should be maximizing after possible optimization. In the USA, more than 80% petroleum wastes are recycled in the refinery where the wastes are generated. Remaining 20% are managing by disposal methods as per local guidelines (Hu et al. 2013).

A variety of methods have been developed and reported since the last few decades including (a) solvent extraction, (b) centrifugation, (c) surfactant enhanced oil recovery, (d) freeze/thaw treatment, (e) sludge pyrolysis, (f) microwave irradiation, (g) electrokinetic methods, and (h) ultrasonic irradiation. These are the major technologies used worldwide for oil recovery from oily sludge. Apart from oil recovery, widely available disposal technologies are incineration, stabilization/solidification, oxidation, and bioremediation. Oil-contaminated soil can be treated through land farming using bio-augmentation, bio-stimulation approach, and bio-pile/composting (Fig. 9.1).

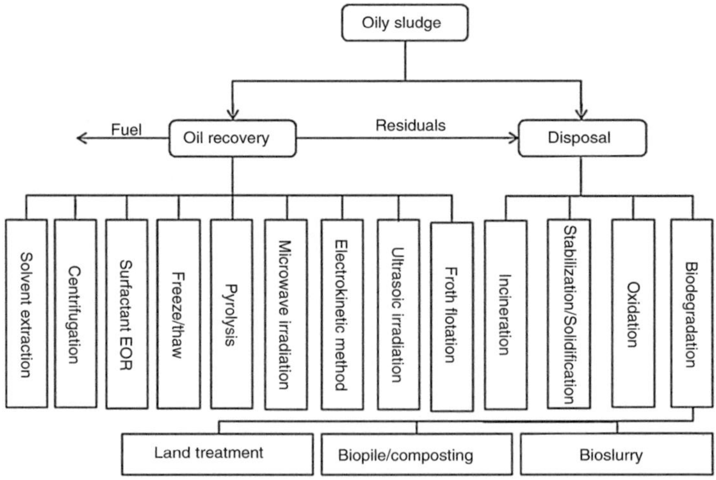

Fig. 9.1 Outline of oily sludge- and oil-contaminated soil treatment technologies

9.3.2 Oily Sludge-/Oil-Contaminated Soil Disposal Technologies

Apart from oil recovery, several technologies are used for disposal of oily sludge including thermal incineration, thermal desorption, stabilization/solidification, oxidation, and biodegradation. The materials recovered after recovery of oil could be directly disposed of to the landfills or apply other technologies if require any further treatment.

9.3.2.1 Incineration Technology

This technology offers a complete combustion of oily sludge in the presence of excess air and auxiliary fuels in a rotary kiln or fluidized bed incinerator. The temperature which ranges from 980 to 1200 °C is used in rotary kiln for 30 min to treat the sludge. The combustion temperature ranges from 730 to 760 °C in the case of fluidized bed incinerator (Scala and Chirone 2004; Hu et al. 2013). Multiple factors are responsible for effective performance of incineration technology including combustion conditions, residence time, temperature, feedstock quality, presence of auxiliary fuels, and waste feed rates (Hu et al. 2013). The advantage of this technology is generation of energy in incinerator which could be used for running steam turbine and significant reduction of the volume of oily sludge after treatment. The major disadvantage is the requirement of pretreatment which is very costly to remove moisture for improvement of combustion efficiency. Apart from that, ash residue scrubber water and scrubber sludge generated during incineration process are very hazardous and need further treatment.

9.3.2.2 Stabilization/Solidification Technology

This technology is considered quick and inexpensive compared to other technologies. Stabilization/solidification enables immobilization of contaminants and transforms to less soluble and less toxic stable form which has no threat to the environment. Inorganic wastes are easily disposed of compared to organic waste by this technology (Faschan et al. 1991; Vipulanandan 1995; Islam 2015). The use of this disposal method for organic waste by rice husk ash as a replacement of cement has been reported (Zain et al. 2010). Li et al. (2015) investigated an innovative approach where oily sludge was treated by modified thermal distillation method where residual asphalt-like emulsion was used as a stabilization/solidification material.

9.3.2.3 Oxidation Treatment Technology

Oxidation technologies are used for degradation of various organic compounds through chemical and other enhanced oxidation process (Islam 2015). The oxidation treatment can be carried out by several oxidants such as Fenton's reagent, hypochlorite, ozone, hydrogen peroxide, permanganate, and persulfate. These compounds generate hydroxyl radicles which react with organic and inorganic compounds and enhance the oxidation process (Ferrarese et al. 2008; Hu et al. 2013). Several groups have proven that combination with other technologies such as sonolysis and ultrasonic irradiation could enhance the remediation process (Zhang et al. 2012; Adewuyi

2001; Dewulf and Langenhove 2001; Peter 2001; Lim et al. 2007). Other enhanced oxidation techniques are available for remediation of oil contamination such as supercritical water oxidation (SCWO), wet air oxidation (WAO), and photocatalytic oxidation (PO) (Hu et al. 2013).

9.3.2.4 Biodegradation and Bioremediation

Biodegradation is commonly a disintegration process of any compound carried out by bacteria, fungi, higher plants, or any other biological means. Bioremediation is a waste management technology that involves the use of microorganisms to remove or neutralize pollutants from a contaminated site. This technology is globally accepted for restoration of oil-polluted environments. Other contaminants like pesticides, heavy metals, etc. could also be mineralized through bioremediation. Several bioremediation technologies are available such as phytoremediation, bioventing, bioleaching, land farming, bioreactor, composting, bio-augmentation, bio-stimulation, and rhizo-filtration. Widely used bioremediation approaches are land farming, bio-pile/composting, and bio-slurry treatment (Hu et al. 2013). Oil industries are continuously generating sludge during operations, and soils beneath the sludge are also contaminated. Many of the constituents from sludge and soil are carcinogenic and immune-toxic. Microorganisms used for bioremediation are naturally occurring, indigenous, or cultivated in the laboratory after selection of the best performer. In situ and ex situ approaches of bioremediation with application of these microorganisms are very common for reclamation of contaminated sites with reduction of the threat of groundwater contamination and enhance the rate of biodegradation (Mishra et al. 2001). However, oily sludge or contaminated soil contains mixture of alkanes, aromatic, NSO (nitrogen-, sulfur-, and oxygen-containing) compounds, and asphaltene fractions. Single bacterial species has only limited capacity to degrade all the fractions of hydrocarbons present in sludge and contaminated soil (Bartha 1986; Bossert and Bartha 1984; Dibble and Bartha 1979; Loser et al. 1998). Indigenous microorganisms in the contaminated soil can degrade a wide range of target constituents of the oily sludge and contaminated soil, but their population and the efficiency are affected when any toxic contaminant is present at higher concentration. Several research had been carried out across the globe and identified single microorganisms and consortia which could degrade oily sludge and oil component fractions (Lal and Khanna 1996a, b; Mishra et al. 2001; Bhattacharya et al. 2003; Sarma et al. 2004, 2010) and tested them at laboratory scale to field scale under tropical to temperate climatic regions (Mandal et al. 2012a, b, c). Few engineered bacteria have also been used for bioremediation of oily sludge. These bacteria showed higher degradative potential in laboratory scale. However, ecological and environmental concerns and regulatory constrains are the major obstacle for testing under field conditions (Das and Chandran 2011). These microbial resources and technologies are used by different companies to treat oil spills and oily sludge across the globe. Nowadays some multinational companies are doing these bioremediation jobs especially in the Middle East countries where huge amount of oily sludge has been generated since more than 50 years during the operation of oil exploration and processing. Now, all local environmental protection authorities are very strict in implementing environmental regulations as per international environmental standards for environmental pollution monitoring. Hence, there is a need for reclamation of oily sludge-contaminated land through bioremediation without any further contamination of environment. Most of the companies are now using their own customized technology based on bioremediation, and they are mainly followed land farming, bio-pile/composting, bio-slurry treatment, and bio-pyramid technologies.

9.4 Recent Developments and Future Prospects

Oil has been used as energy source since many years. The oil and hydrocarbon sector at a global scale has been undergoing radical changes leading to increased industrial activity in the area of hydrocarbon processing. As a result, several oil spills have occurred around the world (Table 9.1). In 2001, the global environmental market, including hazardous waste management and disposal, approaches to brownfield redevelopment, and site remediation, was reported to be of the order of $1 trillion (Masons Water Yearbook 2000–2001). After literature survey, Singh et al. (2009) reported the current status of the international market for remediation sector is estimated to be in the range of USD 30–35 billion. Application of bio-based remediation technologies is rapidly booming according to pre-projection range of USD 1.5 billion per annum (Singh et al. 2009). The soil remediation sector already had a stable market in several developed countries like the USA, Canada, Western European countries, Japan, and Australia. The global remediation market has dynamically exhibited major changes in the Middle East countries since last few years after Gulf War events of 1991 and 2003. The market was growing and refining by identification of new contaminated sites after repeated attack during the Gulf War. Furthermore, public environmental authorities in Middle East countries are more focused on environmental regulations and management to maintain international environmental standards. Several countries prepared their own strategies and action plans to handle these environmental hazards. Kuwait, Lebanon, and many other Middle East countries formed an oil spill working group with the help of environmental activists and environmental research organizations to follow this issue and have undergone an extensive assessment of contamination for affected areas which would contribute to a future strong market in remediation sector.

9.4.1 Kuwait Environmental Remediation Program: A Remediation Case Study

The state of Kuwait is located in the northwest corner of the Arabian Gulf and covers an area of 17,818 km^2, where 114 km^2 of its desert environment were damaged by Iraqi troops through789 detonated oil wells. Crude oil gushed from the damaged oil

Table 9.1 Recent major oil spills worldwide (2010–2016)

Spill/vessel	Location	Dates	Approximate quantity of oil spill (tonnes)
Fox Creek pipeline leak	Alberta, Canada	6th October 2016	213
Colonial Pipeline leak	Alabama, USA	12th September 2016	1092
North Battleford pipeline spill	North Battleford, Canada	21st July 2016	210
ConocoPhillips Canada pipeline spill	Alberta, Canada	9th June 2016	323
Shell Gulf of Mexico oil spill, Brutus offshore platform	Gulf of Mexico, USA	12th May 2016	316
Refugio oil spill	California, USA	19th May 2015	330
MV Marathassa	British Columbia, Canada	13th April 2015	2.3
Yellowstone River oil spill	Montana, USA	17th January 2015	160
Black Sea oil spill	Black Sea, Russia	24th December 2014	Unknown
Trans-Israel Pipeline	Eilat, Israel	6th December 2014	4300
Mid-Valley Pipeline	Louisiana, USA	13th October 2014	546
Lake Michigan oil spill	Indiana, USA	24th March 2014	5
MV Miss Susan/MV Summer Wind	Texas, USA	22nd March 2014	546
North Dakota pipeline spill	North Dakota, USA	21st March 2014	110
North Dakota train collision	North Dakota, USA	30th December 2013	1300
North Dakota pipeline spill	North Dakota, USA	25th September 2013	2810
Lac-Mégantic derailment	Quebec, Canada	6th July, 2013	4830
Mayflower	Mayflower, USA	30th March 2013	950

(continued)

Table 9.1 (continued)

Spill/vessel	Location	Dates	Approximate quantity of oil spill (tonnes)
Arthur Kill storage tank spill (Hurricane Sandy)	New Jersey, USA	29th October 2012	1130
Guarapiche River	Maturin, Venezuela	4th February 2012	41,000
Nigeria	Bonga Field, Nigeria	21st December 2011	5500
Little Buffalo oil spill	Alberta, Canada	29th April 2011	3800
Kalamazoo River oil spill	Michigan, USA	26th July 2010	3250
Xingang Port oil spill	Yellow Sea, China	16th July 2010	90,000
ExxonMobil	Niger Delta, Nigeria	1st May 2010	95,500
Deepwater Horizon	Gulf of Mexico, USA	20th April 2010	6,27,000

wells from ten oil fields forming lakes and contaminated further over 40 km^2 of the land. As a consequence, desert ecosystem was damaged and altered the soil properties. It caused massive death of native plants and animals and penetration of oils to the subsurface layer threatening groundwater contamination (Al-Gharabally and Al-Barood 2015). Apart from Iraqi invasion, several pits including sludge, effluent, and gatch pits are contaminated through recovered oil from any spills, produced water, and construction purposes, respectively, during oil exploration since more than 50 years. In 1991, the United Nation through its security council formed the United Nation Compensation Commission (UNCC) to process claims and pay compensations for loss and damage suffered as a direct result of unlawful Iraqi invasion and occupation of Kuwait during 1990–1991. Very recently, Kuwait National Focal Point and Kuwait Oil Company collaborated a joint project to remediate approximately 26 million cubic meter of heavily contaminated soil through funds provided by UNCC. This project initiated on 2007 as a three-phase project and will continue till 2035–2040 to complete remediation and rehabilitation process. From 2007 to 2009, risk assessment study has been carried out. From 2009 onward remediation and rehabilitation works were initiated on a trial basis at high-risk priority sites. Three bioremediation companies are executing the phase one remediation and rehabilitation across the globe. The Energy and Resources Institute (TERI) from India is one among them whose capability was evaluated by agencies. Subsequently, phase two and phase three of the remediation project will start very

Table 9.2 Major soil remediation companies worldwide

Name	Country
Advanced Future Group International (joint venture With Ivey International INC.)	Canada and Kuwait
Aecom Middle East LTD	Dubai
Biogenesis Enterprises INC	USA
Blackwell Masterton International LTD.	United Kingdom
Clean Harbors Environmental Services, INC.	USA
Coffey International Development (MIDDLE EAST) PTY LTD.	Saudi Arabia
DCI Environmental, INC.	USA
Dekonta, A.S.	Czech Republic
Deme Environmental Contractors NV (DEC NV)	Belgium
Ecophile Company LTD	Korea
Environmental Earth Sciences International	Australia
Environmental Solutions For Petroleum Services – FREE ZONE S.A.E	Egypt
Fomento De ConstruccionesyContratas, S.A (FCC SA)	Spain
Golder Associates S.R.L	Italy
GS Engineering & Construction CORP.	South Korea
Halla Engineering & Construction Corporation	Abu Dhabi
Halliburton Overseas Limited	Kuwait
Hera AG Ambiental S.L.	Spain
Lamor Corporation AB	Finland
LE Floch Depollution	France
Nizhny Novgorod Institute of Applied Technologies, LLC	NizhnyNovgorod The Russian Federation
ONGC TERI Biotech Limited	India
The Energy and Resources Institute	India
SAR A.S.	Norway
Sensatec GMBH	Germany

Ref. from Kuwait Oil Company Web Database

soon for removing the contaminations from medium- to low-risk sites and will continue probably till 2040. More companies from across the world have responded and prequalified the bid to join this bioremediation consortium for removal of huge amount of contaminants (Table 9.2).

9.5 Conclusion

Petroleum contamination is a growing environmental threat, and field-based remediation strategies are being increasingly used globally to tackle oil spill contamination on a large scale. In the past few years, emerging technologies like bioremediation and phytoremediation have become more popular. Bioremediation has begun to compete with other established physical- and chemical-based technologies. The use of bioremediation on a large commercial scale for cleaning up of contamination has increased dramatically since the last two decades and expected to rise further in the next few decades due to more oil spill events associated with continuous exploration and processing of oils across the globe and also continuous turmoil in Middle East countries. Only few companies are available to offer soil remediation through bioremediation technology approaches. Currently, strict environmental regulations are being implemented by all countries to combat global warming and reduced the emission of greenhouse gases. The physical- and chemical-based remediation technologies nowadays are less acceptable to the society. As a consequence, bioremediation approach will become more popular and acceptable to global agencies for removal of these contaminants.

References

Adewuyi YG (2001) Sonochemistry: environmental science and engineering applications. Ind Eng Chem Res 40:4681–4715

Al-Gharabally D, Al-Barood A (2015) Kuwait environmental remediation program (KERP): remediation demonstration strategy. Biological and Chemical Research 2015:289–296

API (1989) API environmental guidance document: onshore solid waste management in exploration and production operations. American Petroleum Institute (API), Washington DC

Bartha R (1986) Biotechnology of petroleum pollutant biodegradation. Microb Ecol 12:155–172

Bastin E (1926) Microorganisms in oilfields. Science 63:21–24

Bhattacharya D, Sarma PM, Krishnan S, Mishra S, Lal B (2003) Evaluation of genetic diversity among *Pseudomonas citronellolis* strains isolated from oily sludge-contaminated sites. Appl Environ Microbiol 69(3):1435–1441

Bhattacharyya JK, Shekdar AV (2003) Treatment and disposal of refinery sludges: Indian scenario. Waste Management Res 21(3):249–261

Bossert I, Bartha R (1984) The fate of petroleum in the soil ecosystems. In: Atlas RM (ed) Petroleum microbiology. Macmillan, New York, pp 435–473

da Silva VL, Alves FC, de Franca FP (2012) A review of the technological solutions for the treatment of oily sludges from petroleum refineries. Waste Manage Res 30:1016–1030

Das N, Chandran P (2011) Microbial degradation of petroleum hydrocarbon contaminants: an overview. Biotechnol Res Int 2011: Article ID 941810, 13 p https://doi.org/10.4061/2011/941810

Dewulf J, Langenhove HV (2001) Ultrasonic degradation of trichloroethylene and chlorobenzene at micromolar concentrations: kinetics and modelling. Ultrason Sonochem 8:143–150

Dibble JT, Bartha R (1979) The effect of environmental parameters on the biodegradation of oily sludge. Appl Environ Microbiol 37:729–739

Elektorowicz M, Habibi S (2005) Sustainable waste management: recovery of fuels from petroleum sludge. Can J Civil Eng 32:164–169

Faschan A, Tittlebaum M, Cartledge F, Eaton H (1991) Effects of additives on solidification of API separator sludge. Environ Monitoring Assess 18:145–161

Ferrarese E, Andreottola G, Oprea IA (2008) Remediation of PAH-contaminated sediments by chemical oxidation. J Hazard Mater 152:128–139

Head IM, Jones DM, Larter SR (2003) Biological activity in the deep subsurface and the origin of heavy oil. Nature 426:344–352

Hu G, Li J, Zeng G (2013) Recent development in the treatment of oily sludge from petroleum industry: a review. J Hazard Mater 261:470–490

Islam B (2015) Petroleum sludge, its treatment and disposal: a review. Int J Chem Sci 13 (4):1584–1602

Krejci-Graf K (1932) Rule of density of oils. Bull Am Assoc Pet Geol 16:1038

Kumar B, Raj Mohan B (2013) Petroleum oily sludge and the prospects of microwave for its remediation. Int J Eng Res Technol 2(11):359–370

Lal B, Khanna S (1996a) Degradation of crude oil by *Acinetobacter calcoaceticus* and *Alcaligenes odorans*. J Appl Bacteriol 81:355–362

Lal B, Khanna S (1996b) Mineralization of ^{14}C Octacosane by *Acinetobacter calcoaceticus* S30. Can J Microbiol 42:1225–1231

Li G, Guo S, Ye H (2015) Thermal treatment of heavy oily sludge: resource recovery and potential utilization of residual asphalt-like emulsion as a stabilization/solidification material. RSC Adv 5:105299–105306

Lim MH, Kim SH, Kim YU, Khim J (2007) Sonolysis of chlorinated compounds in aqueous solutions. Ultrason Sonochem 14:93–98

Loser C, Seidel H, Zehnsdarf A, Stoltmeister U (1998) Microbial degradation of hydrocarbons in soil during aerobic/anaerobic changes and under purely aerobic conditions. Appl Microbiol Biotechnol 49:631–636

Mandal A, Sarma PM, Jeyaseelan CP, Channashettar VA, Singh B, Lal B, Datta J (2012a) Large scale bioremediation of petroleum hydrocarbon contaminated waste at Indian oil refineries: case studies. Int J life Sci Pharm Res 2:114–128

Mandal A, Sarma PM, Singh B, Jeyaseelan CP, Channashettar VA, Lal B, Datta J (2012b) Bioremediation: an environment friendly sustainable biotechnological solution for remediation of petroleum hydrocarbon contaminated waste. IJREISS 2(8):1–18

Mandal A, Sarma PM, Singh B, Jeyaseelan CP, Channashettar VA, Lal B, Datta J (2012c) Bioremediation: an environment friendly sustainable biotechnological solution for remediation of petroleum hydrocarbon contaminated waste. ARPN J Sci Technol 2:1–12

Masons Water Yearbook (2000–2001) http://www.eic-yearbook.co.uk/rem_ind.htm

Mishra S, Jyoti J, Kuhad RC, Lal B (2001) Evaluation of inoculum addition to stimulate in situ bioremediation of oily-sludge-contaminated soil. Appl Environ Microbiol 67(4):1675–1681

Olajire AA (2014) The petroleum industry and environmental challenges. J Pet Environ Biotechnol 5(4). https://doi.org/10.4172/2157-7463.1000186

Peter D (2001) Sonolytic degradation of volatile pollutants in natural ground water: conclusions from a model study. Ultrason Sonochem 8:221–226

Pinheiro BCA, Holanda JNF (2009) Processing of red ceramics incorporated with encapsulated petroleum waste. J Mater Process Technol 209:5606–5610

Roadifer RE (1987) In: Meyer RF (ed) Exploration for heavy crude oil and natural bitumen. American Association of Petroleum Geologists, Tulsa, pp 3–23

Sarma PM, Bhattacharya D, Krishnan S, Lal B (2004) Degradation of polycyclic aromatic hydrocarbons by a newly discovered enteric bacterium, *Leclercia adecarboxylata*. Appl Environ Microbiol 70(5):3163–3166

Sarma PM, Duraja P, Deshpande S, Lal B (2010) Degradation of pyrene by an enteric bacterium, *Leclercia adecarboxylata* PS4040. Biodegradation 21:59–69. https://doi.org/10.1007/s10532-009-9281-z

Scala F, Chirone R (2004) Fluidized bed combustion of alternative solid fuels. Exp Thermal Fluid Sci 28:691–699

Singh A, Kuhad RC, Ward OP (2009) Biological remediation of soil: an overview of global market and available technologies. In: Singh A, Kuhad RC, Ward OP (eds) Advances in applied bioremediation. Springer, New York, pp 1–20

Vipulanandan C (1995) Effect of clays and cement on the solidification/stabilization of phenol-contaminated soils. Waste Manag 15(5/6):399–406

Zain AM, Mahmud H, Shaaban MG (2010) Petroleum sludge stabilization and solidification: an alternative treatment using ordinary Portland cement and rice husk ash. In: 2nd international conference on Chemical, Biological and Environmental Engineering, Cairo, 2010, pp 30–34

Zhang J, Li JB, Thring RW, Hu X, Song XY (2012) Oil recovery from refinery oily sludge via ultrasound and freeze/thaw. J Hazard Mater 203–204:195–203

Microbial Soil Enzymes: Implications in the Maintenance of Rhizosphere Ecosystem and Soil Health

10

Swati Joshi, Balaram Mohapatra, and J. P. N. Mishra

Abstract

Soil enzymes play a crucial role in agriculture as they are important for several vital biochemical reactions necessary for the life processes of soil microbes along with the maintenance of soil structure, decomposition and formation of organic matter and nutrient mineralization. Thus, soil enzymes are instrumental in the maintenance of soil ecosystem. Several factors that affect soil-plant-microorganism and their interaction in turn determine the productivity and activity of soil enzymes. A better understanding of the role of soil enzymes in maintaining soil health, along with the development of rapid and easier protocols for their measurement, will provide us with the opportunity to design novel integrated soil assessment methods leading to more efficient and eco-friendly soil management programmes. Furthermore, harnessing beneficial soil enzymes through currently available technologies will enable in situ applications of desired soil enzymes for restoration of polluted soils. Present article focuses on occurrence of soil enzymes, methods for determining their activity, current developments in mining these enzymes using metagenomic approach and factors affecting the activity of soil enzymes along with the importance of various soil enzymes.

S. Joshi (✉)
School of Life Sciences, Central University of Gujarat, Gandhinagar, Gujarat, India
e-mail: swati.joshi.aj@gmail.com

B. Mohapatra
Environmental Microbiology Laboratory, Indian Institute of Technology, Kharagpur, West Bengal, India

J. P. N. Mishra
School of Life Sciences, Central Univeristy of Gujarat, Gandhinagar, Gujarat, India

T. K. Adhya et al. (eds.), *Advances in Soil Microbiology: Recent Trends and Future Prospects*, Microorganisms for Sustainability 3,
https://doi.org/10.1007/978-981-10-6178-3_10

Keywords
Microbial soil enzymes · Soil ecosystem · Rhizosphere · Soil health · Metagenomics

10.1 Introduction

Soil is a crucial part of the terrestrial ecosystem and provides fundamental support to all terrestrial life forms. Therefore, proper soil protection programmes are indispensable to avoid problems of soil erosion, infertility, contamination of groundwater and poor water holding capacity and most importantly to avoid the loss of biodiversity. Soil quality is directly related to biological properties of the soil, which are very sensitive to any environmental disturbances. Soil microbiota is equally sensitive and changes quickly in response to environmental perturbations. Profile of soil microbiota and enzymes is interrelated and an important indicator of soil health and quality (Pajares et al. 2011). Soil enzymes which are mainly produced by soil-inhabiting microbes play a crucial role in nutrient cycling and reflect soil microbial activity and fertility (Bentez et al. 2000). A fine balance of biological (which includes enzymatic activity), chemical and physical components is essential to maintain soil health. Soil enzymes play a key role in the overall process of decomposition of organic matter in the soil (Sinsabaugh et al. 1991). Thus, it is clear that a balanced soil enzyme system is imperative to maintain soil processes.

Rhizosphere region of the soil is biologically very active and rich in soil enzymes as compared to rest of the soil. Functions of the plant roots modify biogeochemical parameters of the soil and bring changes in nutrient levels, pollutants, concentrations of various chelating compound, pH and redox potential, partial pressures of oxygen and carbon dioxide [pO_2 and pCO_2], etc. (Gianfreda 2015). Production of soil enzymes in the rhizosphere region and their activity depends on several factors. A higher rhizosphere enzyme activity is correlated to greater functional diversity of resident microbial population. Absence or inhibition of soil enzyme activities reduces processes that can result in poor nutrition of the plants. Suppression of certain enzyme activity (e.g. pesticide-degrading enzymes) can result in pile-up of harmful chemicals in the soil; some of these chemicals may further inhibit soil enzyme and deteriorate soil quality. Although there are several publications on soil enzymes piling up day by day, several key questions are yet not answered. In this chapter, we try to shed light on key aspects related to soil enzymology, in situ and ex situ estimation of soil enzymes, factors affecting their activity and molecular approach to mine soil enzyme-encoding genes and significance of soil enzymes.

10.2 Sources of Soil Enzymes

Sources of soil enzyme mainly include microbes, plant roots and soil animals. Since long soil enzymes have been categorized in two main groups, extracellular soil enzymes and intracellular soil enzymes (Burns et al. 2013), enzymes that are present and function inside the living cell are grouped under the latter category, while those produced by living cells but secreted outside comprise the latter. In long terms, soil enzymes get stabilized by accumulation and complexation with humus (organic matter) in the soil. These stabilized soil enzymes, which are no longer associated with viable cells, contribute to about 40–60% activity of soil enzymes (http://soilquality.org/home.html). An exception to this is dehydrogenases, which can only be produced by living cells, thus contributed to the pool of soil enzymes by viable cells only (Yuan and Yue 2012). Various biochemical, chemical and physiochemical reactions play their role in carrying out nutrient cycles in soil. Extracellular enzymes help decay of organic matter of the soil and aid in mineralization of soil organic carbon (C), phosphorus (P) and nitrogen (N) (Bandick and Dick 1999; Finzi et al. 2006). Forest soils play a significant role in the global carbon cycle (Jobággy and Jackson 2000).

Soil microbe mediates biochemical processes along with plant roots and soil animals. Biochemical processes are carried out by a host of enzymes (β-glucosidase, α-glucosidase, β-xylosidase, amylase, chitinase, dehydrogenase, urease, protease, phenol oxidase, L-leucine aminopeptidase, *N*-acetyl-glucosaminidase, phosphatase, arylsulphatases) that are found in soil (Miwa et al. 1937; Zahir et al. 2001; Ji et al. 2014; Herold et al. 2014). Bulk of the enzyme activity is contributed by microbes that can be rationalized by their large biomass, comparatively higher metabolic activity and larger quantities of secretion of extracellular enzymes into the soil solution (Spier and Ross 1978). Production of several polymer degrading enzymes is commonly ascribed to fungi (Hättenschwiler et al. 2005; Baldrian and Valášková 2008). Saprotrophic species of *Basidiomycota* are known to be the exclusive producers of ligninolytic enzymes such as Mn-peroxidase and lignin peroxidase (Hofrichter 2002; Baldrian and Valášková 2008). Arylsulphatases are widespread in soils (Dodgson et al. 1982; Gupta et al. 1993; Ganeshamurthy et al. 1995). Primarily they are secreted by bacteria into the external environment in response to sulphur limitation and hydrolyze sulphate esters in soil (McGill and Colle 1981; Kertesz and Mirleau 2004). So far, very limited information is available regarding arylsulphatases synthesizing specific microbial genera that play significant role in the soil organic sulphur cycle (Kertesz and Mirleau 2004). Chitinase, which hydrolyze chitin (poly β-1-4-(2-ncetamido-2-deoxy)-D-glucoside) is an agriculturally important class of soil enzymes. These are produced by both microbes and plants and have been reported to control various soil-borne diseases by hydrolysing the cell wall of phytopathogenic fungi such as *Sclerotium rolfsii* and *Rhizoctonia solani* (Ordentlich et al. 1988; Shapira et al. 1989). Supplementation of chitinase to frequently applied chemical fungicides will not only make them effective but also minimize the use of otherwise harmful chemical insecticides and fungicide, contributing to sustainable agriculture (Gunaratna and Balasubramanian 1994; Wang et al. 2002). Phosphatases are abundant in rhizospheric region as compared to the bulk soil and exhibit a very good relationship with mycorrhizal association (Kumar et al. 2011). Wu et al. (2012) studied protease and β-glucosidase in the rhizosphere region of *Citrus unshiu* and established a correlation of these enzymes with root mycorrhiza, spatial distribution of glomalin-related soil proteins (GRSP) and carbohydrates. Similarly, in a fire chronosequence in Alaska, Gartner et al.

(2012) found a correlation between five enzymes involved in the transformation of C, P and N substrates and in the presence of mycorrhiza. Thus, mycorrhiza has also been an important source of soil enzymes along with other microbial sources.

10.3 Methods to Determine Soil Enzyme Activities

Enzyme activities in soil affect various aspects of soil biology and are very useful for gauging soil fertility, functional diversity of soil microbiota and overall turnover of organic compounds in soil systems at different geographical locations (Kandeler et al. 1999). Estimation of the soil enzymes remained a challenge for several years due to want of appropriate quantitative and qualitative techniques. However, recent advances in the field of soil enzymology have enabled us to measure the soil enzyme activity both in situ and ex situ assay. Assay methods that provide reliable results on soil enzyme concentration and rate of the reaction have also been developed (Baldrian 2009). Ndiaye et al. (2000) observed that any change in soil management approach and land use technique results in corresponding changes in the soil enzyme activities and suggested that alterations in soil quality can be anticipated by recording changes in soil enzyme profile, before they are detected by any other soil analyses methods. As discussed by Rao et al. (2014), currently available methods to assay soil enzyme activity suffer with several limitations:

(a) These methods do not provide adequate information on real enzyme activities but measure the potential enzyme activities.
(b) They take into consideration and provide information on stabilized enzymes which might not be active at conditions prevailing under in situ soil environment.
(c) They do not furnish any information related to production and origin of the soil enzyme
(d) They do not provide information on changes occurring in enzyme activity that occur in continuously changing in situ environment conditions.
(e) As soil enzymes are part of complex and dynamic processes, estimation of single enzyme activity provides no clue about their role in such dynamics.
(f) In soil, enzymes are exposed to several environmental, physicochemical, anthropogenic activities, and laboratory assays do not allow correct interpretation of the effect of such disturbances on soil enzyme activity.

In case of in situ assay of soil enzymes, a different set of technological challenges exists including hindrances by electron-dense humic substances and soil minerals and reduced rate of diffusion which reduces overall on-site interaction of enzymes and substrate (Steinweg et al. 2012). Methods including electron microscopic observation (Ladd et al. 1996), zymography (Spohn et al. 2013) and near-infrared spectroscopy (Dick et al. 2013) have been employed to estimate the enzyme activity directly on site. Majority of researchers opt for assays which are carried out under laboratory conditions. In such assays it is imperative to give careful details of soil sampling, handling, storage and enzyme assay so that the method can be reproduced and compared with other studies. Also, extraction of the enzymes is carried out before performing the biochemical assay. A considerable amount of the enzymes is bound to soil components or microbial biomass and is not extractable and thus remains out of the estimation (Claus and Filip 1990; Valásková and Baldrian 2006).

As compared to organic matter-rich forest soils, enzyme extraction from high clay-containing soils is found to be poor (Vepsäläinen 2001; Šnajdr et al. 2008b). Further processing of the extracts is required to get rid of inhibitory compounds such as heavy metals and humic acid (Baldrian and Gabriel 2002; Zavarzina et al. 2004). Vancov and Keen (2009) developed a rapid and high-throughput method of enzyme extraction from soil. They reported that their 1-day extraction protocol included physical disruption of the soil samples with bead beating and was reproducible. In a study carried out by DeForest (2009), it was clearly demonstrated that soil storage conditions and processing method significantly affect the estimation of enzymatic activity in acidic soil of forest. In this study, six extracellular enzymes were measured [employing 4-4-methylumbelliferone (MUF)-linked substrates and L-dihydroxyphenylalanine (L-DOPA)] from soil samples stored for varying time duration at different temperatures. Results of this study revealed that in contrast to storage temperature, enzyme activity values were affected by extended time in buffer. It has been observed that freezing of the soil sample affect soil enzyme activity more than air-drying of the sample (Wallenius et al. 2010; Peoples and Koide 2012). Fluorimetric and spectrophotometric assays [which employ p-nitrophenol (pNP)- and MUF(4-4-methylumbelliferone)-based substrates] are very popular and routinely used for measuring activity of soil hydrolases such as glucosaminidase, glucosidase, galactosidase, etc. (Moscatelli et al. 2012; Trap et al. 2012; Dick et al. 2013). For high-throughput results, these assays are also being carried out by using microplate methods (Trap et al. 2012).

10.4 Factors Affecting Soil Enzymes

Soil enzyme activities are very sensitive to any external disturbances including both anthropogenic and climatic perturbations (Vepsäläinen 2001). Several physicochemical and biological factors affect either enzyme quantities or their activity levels. For example, enzyme activity in soils changes with seasonal variables in moisture, temperature and addition of fresh litter. Like other enzymes, soil enzymes also exhibit varying optimum pH and temperature at which they are most active. For instance, activity of arylsulphatase, phosphatase and amidase involved in sulphur, phosphorus and nitrogen cycling, respectively, is strongly correlated to alteration in pH of the soil (Tabatabai 1994; Kertesz and Mirleau 2004; Chaudhari and Bhatt 2014). Temperature affects several aspects of soil enzymes such as soil enzyme activities, stability and enzyme kinetics, substrate affinity and production levels of enzyme as it also influences the activity and population of soil microbes (Wallenstein et al. 2009; Baldrian et al. 2013). Heat and extreme cold temperature can alter enzyme structure and substrate binding site and therefore, can decrease the enzyme activity above and below the temperature optimum. In a study carried out by McClaugherty and Linkins (1990), it was observed that there was 33–80% decline in the chitinase, peroxidase and laccase activities in winter samples where temperature remains at 0 °C, as compared to those in autumn samples where temperature reaches 15 °C. This is evidence that the seasonal patterns of temperature of ecosystems can affect activity of soil enzymes. The activity of many enzymes often correlates with soil moisture content, as well. Drought may suppress enzyme activity (Sardans and Penuelas 2005; Gömöryová et al. 2006; Baldrian et al. 2010). Upon reduction of 21 % of soil moisture, a corresponding reduction in urease, protease, β-glucosidase and acid phosphatase activity was recorded by Sardans and Penuelas (2005).

Levels of organic matter, nitrogen content and various macronutrients appear to regulate the production of enzymes in soil. It has been reported that with increase in organic matter content, activity of several hydrolytic enzymes including cellobiohydrolase, β-glucosidase, phosphatase and N-acetylglucosaminidase increases (Nsabimana et al. 2004; Sinsabaugh et al. 2008). Decrease in the levels of N-acetylglucosamine liberating enzyme chitinase has been associated with increasing nitrogen content in the soil environment (Olander and Vitousek 2000; Andersson et al. 2004). According to Prietzel (2001), addition of $(NH_4)_2SO_4$ reduces activity of arylsulphatase. Addition of nitrogen can significantly affect the kinetics of the soil enzyme as observed in case of β-glucosidase, β-xylosidase, cellobiohydrolase and β-N-acetylglucosaminidase involved in soil organic matter degradation in forest soils (Stone et al. 2012). In general, the presence of available phosphorous (P) in soil is related to decreases in phosphatase activity (Venkatesan and Senthurpandian 2006). It has also been confirmed by many researchers across the globe that the presence of pollutants such as heavy metals and organic xenobiotic changes the enzyme profile of the soil (Burns and Dick 2002; Effron et al. 2004).

Although soil enzymes of microbial origin are produced by a diverse array of microbes, production of certain enzymes are limited to certain taxa. Fungi are the most common producer of lignocellulose-hydrolysing enzymes (Moller et al. 1999; Caldwell 2005; Baldrian and Valášková 2008). Chitinase activity is also associated to fungal biomass (Miller et al. 1998; Sinsabaugh et al. 2008). Through microcosm studies it has been demonstrated that introduction of saprophytic fungi in soil increases the activity of different oxidative and hydrolytic enzymes (Šnajdr et al. 2008a, 2011). In forests, soil enzyme activities vary with the change in the dominant tree species of the forest as it changes the litter input (Weand et al. 2010). At the harvesting time, reduction in microbial population, litter input and alteration in soil microenvironment causes decline in enzyme activity (Hassett and Zak 2005)

Spatial heterogeneity is one of the key attributes of the soil environment (Paul 2007). Changes in enzyme activity have been observed with changes in depth. In forest soil, vertical gradient of enzyme activity is more prominent than any other ecosystem. Fresh carbon input in the form of leaf litter and root exudates makes the surface soil horizons (~ 10 cm thick) carbon rich. Organic compounds thus entering the soil accelerate the growth of microorganisms, which in turn produces extracellular enzymes. However, despite smaller carbon inputs, a significant amount of carbon is also stored in subsoil horizons because of its larger thickness (Wang et al. 2010).

In case of grasslands where extensive root system of trees is absent and agricultural soils where soil homogenization is a routine practice, considerable level of spatial variability with respect to activity of extracellular enzymes and soil chemistry has been observed (Štursová and Baldrian 2011).

Soil type and texture also influence the enzymatic activities. According to Burns (1982), soil texture plays a significant role in stabilizing soil enzymes; importantly the interactions with soil organic matter and clay minerals affect the stability of the enzymes. Studies on soils from different regions have shown that activities of soil enzyme are sensitive to changes in occurring to the soil because of tillage, cropping system and land use (Staben et al. 1997; Gewin et al. 1999; Ndiaye et al. 2000; Acosta-Martinez and Tabatabai 2001; Ekenler and Tabatabai 2002; Ji et al. 2014).

10.5 Mining of Soil Enzymes Encoding Genes Through Metagenomics and Metatranscriptomics

Among soil-inhabiting microbes, a large number of them remain unculturable, but they do contribute to enzyme repertoire of the soil (Lorenz and Eck 2005). The soil metagenome, the collective microbial genome, could be cloned and sequenced directly from soils to search for novel microbial resources. Metagenome analysis has become a remarkable tool to tap yet uncultured microbial diversity present in soil. Recent advances in molecular methods have enabled us to target the abundance of genes encoding enzymes using metagenome or metatranscriptome analysis. These techniques have very high theoretical potential and been employed for assigning gene sequences to specific groups of soil microorganisms along with specifically targeting exocellulase (Baldrian et al. 2012), laccase (Luis et al. 2005; Hassett et al. 2009; Lauber et al. 2009) or a range of various oxidases and glycosyl hydrolases (GH) present in forest soils (Kellner and Vandenbol 2010). The limitations of single-gene surveys (which are applicable for only highly similar gene sequences) might be overcome by sequencing whole transcriptomes that would help the analysis of the entire spectrum of expressed genes. In a recent study by Damon et al. (2012), several families of GHs and other hydrolytic enzymes were detected upon analysis of eukaryotic gene expression in forest soils. Cellulolytic and cell wall-degrading enzymes are of special interest owing to their application in biotechnology sector for bioenergy production. Liu et al. (2011) isolated a low-temperature active, thermostable, halotolerant cellulase from red soil metagenome. Similarly, Verma et al. (2013) fished out a novel thermo-alkali-stable xylanase from compost soil metagenome. Faoro et al. (2012) reported isolation of lipolytic enzyme from forest soil (Paraná state, Brazil). From the mountain soil of north-western Himalayas, Sharma and co-workers (2010) reported the recovery of a cold-adapted amylolytic enzyme. Various other enzymes including oxidases, reductase, racemase, lactonase, esterase, glucosidase, etc. have been isolated from soil metagenome as reviewed by Lee and Lee (2013). Along with metagenomics and metatranscriptomics, a more challenging environmental proteomics approach has demonstrated its potential for analysis of protein pool in soil environment (Schneider et al. 2012). In the near future, it may be one of the most powerful approaches in soil enzyme research.

10.6 Importance of Soil Enzymes

Soil enzymes are of immense significance in maintaining ecophysiological life of soil. Enzyme activity ratios have been employed to analyse ecoenzymatic stoichiometry of freshwater sediments and terrestrial soils (Sinsabaugh et al. 2009), along with the studies on the effects of climate and soil properties of different ecosystems (Sinsabaugh et al. 2008; McDaniel et al. 2013). The ratio of activities of extracellular enzymes, which are related to energy and nutrient acquisition, i.e. ratio of β-glucosidase activity/phosphatase activity (an indicator of potential C/P utilization activity), can be utilized to follow the shifts and pattern of energy supply and demand (Sinsabaugh et al. 2008; McDaniel et al. 2013).

By assessing activities of hydrolases, valuable information can be obtained on the status of key reactions involved in the rate-limiting steps of organic matter decomposition along with those of nutrient transformation. Thus, information on the soil degradation potential can be obtained by knowing soil enzyme activities (Trasar-Cepeda et al. 2000). Bolton et al. (1985) proposed that concomitant estimation of different enzyme activities can be used as an effective indicator of soil microbial activity.

There is an exponentially growing interest in finding and developing green technologies for partial or total recovery of sites with polluted soil. Co-occurrence of different types of polluting compounds (both inorganic and organic) makes the remediation of such sites very problematic. Enzymes can be applied to a large array of different compounds, as enzymes with both narrow (chemo-, region- and stereoselectivity) and broad specificity are known, and therefore can be used in a case-specific manner for transformation of innocuous compounds. Enzymes either released by the plants or by soil microorganisms in rhizosphere and the bulk part of the soil are capable of degrading pollutant reaching to the soil. The representative enzymatic classes in the restoration of polluted environments are hydrolases, dehalogenases, oxidoreductases and transferases. Primarily oxidoreductases and hydrolases effectively degrade and transform phenols, polycyclic aromatic hydrocarbons (PAHs), polychlorinated biphenyls (PCBs) and estrogenic chemicals (Gianfreda and Rao 2004; Gianfreda and Ruggiero 2006). Exploiting the fact that soil enzymes are capable of biodegradation and remediation of xenobiotic compounds, a number of transgenic plants expressing/secreting relevant enzymes have been generated and have been used for restoration of highly polluted soils in various locations (Abhilash et al. 2009). A surge in the release of superoxide dismutase (SOD), ascorbate peroxidase (APX) and catalase (CAT) activities was recorded by the plants when subjected to heavy metal (Hg) pollution of the soil (Li et al. 2013). This observation indicates towards the adaptation of the plant to Hg stress by means of enhanced release of enzymes to deal with the metal stress. Several oxidative enzymes, laccases, catechol dioxygenase, tyrosinase, manganese peroxidase, chloroperoxidase, etc., have been employed in remediation of contaminated soil environments (Duran and Esposito 2000).

10.7 Conclusions

Soil enzymes are of paramount importance for achieving and maintaining physico-chemical and biological balance for soil health. Despite several studies, a universal and accurate methodology is still needed to quantify soil enzymes. While substantial progress has been made towards unravelling soil enzymes, applications of enzymes in soil management programmes are still in its infancy. An integrated approach of discovery, quantification and application has to be developed, so that the potential of soil enzymes can be fully utilized for both environment restoration and human welfare.

References

Abhilash PC, Jamil S, Singh N (2009) Transgenic plants for enhanced biodegradation and phytoremediation of organic xenobiotics. Biotechnol Adv 27:474–488. https://doi.org/10.1016/j.biotechadv.2009.04.002

Acosta-Martı´nez V, Tabatabai MA (2001) Tillage and residue management effects on arylamidase activity in soils. Biol Fertil Soils 34:21–24

Andersson M, Kjřller A, Struwe S (2004) Microbial enzyme activities in leaf litter, humus and mineral soil layers of European forests. Soil Biol Biochem 36:1527–1537

Baldrian P (2009) Microbial enzyme-catalyzed processes in soils and their analysis. Plant Soil Environ 55(9):370–378

Baldrian P, Gabriel J (2002) Copper and cadmium increase laccase activity in *Pleurotus ostreatus*. FEMS Microbiol Lett 206:69–74

Baldrian P, Valášková V (2008) Degradation of cellulose by basidiomycetous fungi. FEMS Microbiol Rev 32:501–521. https://doi.org/10.1111/j.1574-6976.2008.00106.x

Baldrian P, Merhautová V, Cajthaml T et al (2010) Small-scale distribution of extracellular enzymes, fungal, and bacterial biomass in *Quercus petraea* forest topsoil. Biol Fertil Soils 46:717–726. https://doi.org/10.1007/s00374-010-0478-4

Baldrian P, Kolařík M, Štursová M et al (2012) Active and total microbial communities in forest soil are largely different and highly stratified during decomposition. ISME J 6:248–258. https://doi.org/10.1038/ismej.2011.95

Baldrian P, Šnajdr J, Merhautová V et al (2013) Responses of the extracellular enzyme activities in hardwood forest to soil temperature and seasonality and the potential effects of climate change. Soil Biol Biochem 56:60–68. https://doi.org/10.1016/j.soilbio.2012.01.020

Bandick AK, Dick RP (1999) Field management effects on soil enzyme activities. Soil Biol Biochem 31:1471–1479. https://doi.org/10.1016/S0038-0717(99)00051-6

Bentez E, Melgar R, Sainz H et al (2000) Enzyme activities in the rhizosphere of pepper (*Capsicum annuum* L.) grown with olive cake mulches. Soil Biol Biochem 32:1829–1835. https://doi.org/10.1016/S0038-0717(00)00156-5

Bolton H, Elliott LF, Papendick RI et al (1985) Soil microbial biomass and selected soil enzyme activities–effect of fertilization and cropping practices. Soil Biol Biochem 17:297–302. https://doi.org/10.1016/0038-0717(85)90064-1

Burns RG (1982) Enzyme activity in soil: location and a possible role in microbial ecology. Soil Biol Biochem 14:423–427. https://doi.org/10.1016/0038-0717(82)90099-2

Burns RG, Dick RP (2002) Enzymes in the environment. Marcel Dekker, New York, p 614
Burns RG, DeForest JL, Marxsen J et al (2013) Soil enzymes in a changing environment: current knowledge and future directions. Soil Biol Biochem 58:216–234. https://doi.org/10.1016/j.soilbio.2012.11.009
Caldwell BA (2005) Enzyme activities as a component of soil biodiversity: a review. Pedobiologia 49:637–644. https://doi.org/10.1016/j.pedobi.2005.06.003
Chaudhari RG, Bhatt SA (2014) Profiling enzymes involved in nitrogen transformation in semi arid soil. Int J Curr Microbiol App Sci 3(10):1008–1014
Claus H, Filip Z (1990) Effects of clays and other solids on the activity of phenoloxidases produced by some fungi and actinomycetes. Soil Biol Biochem 22:483–488. https://doi.org/10.1016/0038-0717(90)90182-Y
Damon C, Lehembre F, Oger-Desfeux C et al (2012) Metatranscriptomics reveals the diversity of genes expressed by eukaryotes in forest soils. PLoS One 7:E28967. https://doi.org/10.1371/journal.pone.0028967
DeForest JL (2009) The influence of time, storage temperature, and substrate age on potential soil enzyme activity in acidic forest soils using MUB-linked substrates and L-DOPA. Soil Biol Biochem 41:1180–1186. https://doi.org/10.1016/j.soilbio.2009.02.029
Dick WA, Thavamani B, Conley S et al (2013) Prediction of β-glucosidase and β-glucosaminidase activities, soil organic C, and amino sugar N in a diverse population of soils using near infrared reflectance spectroscopy. Soil Biol Biochem 56:99–104. https://doi.org/10.1016/j.soilbio.2012.04.003
Dodgson KS, White G, Fitzgerald JW (1982) Sulphatase enzyme of microbial origin. Afr J Biotechnol Vol I:156–159
Duran N, Esposito E (2000) Potential applications of oxidative enzymes and phenoloxidase-like compounds in wastewater and soil treatment: a review. Appl Catal B Enzym 28:83–99. https://doi.org/10.1016/S0926-3373(00)00168-5
Effron D, de la Horra AM, Defrieri RL et al (2004) Effect of cadmium, copper, and lead on different enzyme activities in a native forest soil. Commun Soil Sci Plant Anal 35:1309–1321. https://doi.org/10.1081/CSS-120037548
Ekenler M, Tabatabai MA (2002) β-Glucosaminidase activity of soils: effect of cropping systems and its relationship to nitrogen mineralization. Biol Fertil Soils 36:367–376. https://doi.org/10.1007/s00374-002-0541-x
Faoro H, Glogauer A, Couto GH et al (2012) Characterization of a new Acidobacteria-derived moderately thermostable lipase from a Brazilian Atlantic Forest soil metagenome. FEMS Microbiol Ecol 81:86–394. https://doi.org/10.1111/j.1574-6941.2012.01361.x
Finzi AC, Sinsabaugh RL, Long TM et al (2006) Microbial community responses to atmospheric carbon dioxide enrichment in a warm-temperate for-est. Ecosystems 9:215–226. https://doi.org/10.1007/s10021-005-0078-6
Ganeshamurthy AM, Singh G, Singh NT (1995) Sulphur status and response of rice to sulphur on some soils of Andaman and Nicobar Islands. J Indian Soc Soil Sci 43:637–641
Gartner TB, Treseder KK, Malcolm GM et al (2012) Extracellular enzyme activity in the mycorrhizospheres of a boreal fire chronosequence. Pedobiologia 55:121–127. https://doi.org/10.1016/j.pedobi.2011.12.003
Gewin VL, Kennedy AC, Veseth R et al (1999) Soil quality changes in eastern Washington with Conservation Reserve Program (CRP) take out. J Soil Water Conserv 54:432–438
Gianfreda L (2015) Enzymes of importance to rhizosphere processes. J Soil Sci Plant Nutr 15 (2):283–306
Gianfreda L, Rao MA (2004) Potential of extra cellular enzymes in remediation of polluted soils: a review. Enzym Microb Technol 35:339–354. https://doi.org/10.1016/j.enzmictec.2004.05.006
Gianfreda L, Ruggiero P (2006) Enzyme activities in soil. In: Nannipieri P, Smalla K (eds) Nucleic acids and proteins in soil, vol 8. Springer, Berlin, pp 257–311
Gömöryová E, Gregor J, Pichler V et al (2006) Spatial patterns of soil microbial characteristics and soil moisture in a natural beech forest. Biologia 61:S329–S333. https://doi.org/10.2478/s11756-006-0183-9

Gunaratna KR, Balasubramanian R (1994) Partial purification and properties of extracellular chitinase produced by *Acremonium obclavatum*, an antagonist to the groundnut rust, *Puccinia arachidis*. World J Microbiol Biotechnol 10(3):342–345. https://doi.org/10.1007/BF00414876

Gupta VVSR, Farrell RE, Germida JJ (1993) Activity of arylsulfatases in Saskatchewan soils. Can J Soil Sci 73:341–347

Hassett JE, Zak DR (2005) Aspen harvest intensity decreases microbial biomass, extracellular enzyme activity, and soil nitrogen cycling. Soil Sci Soc Am J 69:227–235. https://doi.org/10.2136/sssaj2005.0227

Hassett JE, Zak DR, Blackwood CB et al (2009) Are basidiomycete laccase gene abundance and composition related to reduced lignolytic activity under elevated atmospheric NO_3^- deposition in a northern hardwood forest? Microb Ecol 57:728–739. https://doi.org/10.1007/s00248-008-9440-5

Hättenschwiler S, Tiunov AV, Scheu S (2005) Biodiversity and litter decomposition in terrestrial ecosystems. Annu Rev Ecol Evol Syst 36:191–218. https://doi.org/10.1146/annurev.ecolsys.36.112904.151932

Herold N, Schöning I, Berner D et al (2014) Vertical gradients of potential enzyme activities in soil profiles of European beech, Norway spruce and Scots pine dominated forest sites. Pedobiologia 57(2014):181–189. https://doi.org/10.1016/j.pedobi.2014.03.003

Hofrichter M (2002) Review: lignin conversion by manganese peroxidase (MnP). Enzym Microb Technol 30:454–466. https://doi.org/10.1016/S0141-0229(01)00528-2

Ji B, Hu H, Zhao Y et al (2014) Effects of deep tillage and straw returning on soil microorganism and enzyme activities. Sci World J 2014:451493, 12 pages. https://doi.org/10.1155/2014/451493

Jobággy EG, Jackson RB (2000) The vertical distribution of soil organic carbon and its relation to climate and vegetation. Ecol Appl 10:423–436. https://doi.org/10.1890/1051-0761(2000)010[0423:TVDOSO]2.0.CO;2

Kandeler E, Tscherko D, Spiegel H (1999) Long-term monitoring of microbial biomass, N mineralisation and enzyme activities of a Chernozem under different tillage management. Biol Fertil Soils 28:343–351. https://doi.org/10.1007/s003740050502

Kellner H, Vandenbol M (2010) Fungi unearthed: transcripts encoding lignocellulolytic and chitinolytic enzymes in forest soil. PLoS One 5:E10971. https://doi.org/10.1371/journal.pone.0010971

Kertesz MA, Mirleau P (2004) The role of soil microbes in plant sulphur nutrition. J Exp Bot 55:1939–1945. https://doi.org/10.1093/jxb/erh176

Kumar S, Chaudhuri S, Maiti SK (2011) Phosphatase activity in natural and mined soil – a review. Indian J Environ Prot 31:955–962. https://doi.org/10.5829/idosi.mejsr.2013.13.7.2801

Ladd JN, Forster RC, Nannipieri P et al (1996) Soil structure and biological activity. In: Stotzky G, Bollag J-M (eds) Soil biochemistry. Marcel Dekker, New York, pp 23–78

Lauber CL, Sinsabaugh RL, Zak DR (2009) Laccase gene composition and relative abundance in oak forest soil is not affected by short-term nitrogen fertilization. Microb Ecol 57:50–57. https://doi.org/10.1007/s00248-008-9437-0

Lee MH, Lee SW (2013) Bioprospecting potential of the soil metagenome: novel enzymes and bioactivities. Genomics Inform 11(3):114–120. https://doi.org/10.5808/GI.2013.11.3.114

Li Y, Sun H, Li H et al (2013) Dynamic changes of rhizosphere properties and antioxidant enzyme responses of wheat plants (*Triticum aestivum* L.) Chemosphere 93:972–977. https://doi.org/10.1016/j.chemosphere.2013.05.063.

Liu J, Liu WD, Zhao XL et al (2011) Cloning and functional characterization of a novel endo-β-1,4-glucanase gene from a soil-derived metagenomic library. Appl Microbiol Biotechnol 89:1083–1092. https://doi.org/10.1007/s00253-010-2828-4

Lorenz P, Eck J (2005) Metagenomics and industrial applications. Nat Rev Microbiol 3:510–516. https://doi.org/10.1038/nrmicro1161

Luis P, Kellner H, Martin F et al (2005) A molecular method to evaluate basidiomycete laccase gene expression in forest soils. Geoderma 128:18–27. https://doi.org/10.1016/j.geoderma.2004.12.023

McClaugherty CA, Linkins AE (1990) Temperature responses of enzymes in two forest soils. Soil Biol Biochem 22:29–33. https://doi.org/10.1016/0038-0717(90)90056-6

McDaniel MD, Kaye JP, Kaye MW (2013) Increased temperature and precipitation had limited effects on soil extracellular enzyme activities in a post-harvest forest. Soil Biol Biochem 56:90–98. https://doi.org/10.1016/j.soilbio.2012.02.026

McGill WB, Colle CV (1981) Comparative aspects of cycling of organic C, N, S and P through soil organic matter. Geoderma 26:267–286. https://doi.org/10.1016/0016-7061(81)90024-0

Miller M, Palojarvi A, Rangger A et al (1998) The use of fluorogenic substrates to measure fungal presence and activity in soil. Appl Environ Microbiol 64:613–617

Miwa T, Ceng CT, Fujisaki M et al (1937) Zur Frage der Spezifitat der Glykosidasen. I. Verhalted vonβ-d-glucosidases verschiedener Herkunft gegenuberden β-d-Glucosiden mitverschiedenen Aglykonen. Acta Phytochim (Tokyo) 10:155–170

Moller J, Miller M, Kjoller A (1999) Fungal–bacterial interaction on beech leaves: influence on decomposition and dissolved organic carbon quality. Soil Biol Biochem 31:367–374. https://doi.org/10.1016/S0038-0717(98)00138-2

Moscatelli MC, Lagomarsino A, Garzillo AMV et al (2012) β-Glucosidase kinetic parameters as indicators of soil quality under conventional and organic cropping systems applying two analytical approaches. Ecol Indic 13:322–327. https://doi.org/10.1016/j.ecolind.2011.06.031

Ndiaye EL, Sandeno JM, McGrath D et al (2000) Integrative biological indicators for detecting change in soil quality. Am J Altern Agric 15:26–36. https://doi.org/10.1017/S0889189300008432

Nsabimana D, Haynes RJ, Wallis FM (2004) Size, activity and catabolic diversity of the soil microbial biomass as affected by land use. Appl Soil Ecol 26:81–92. https://doi.org/10.1016/j.apsoil.2003.12.005

Olander LP, Vitousek PM (2000) Regulation of soil phosphatase and chitinase activity by N and P availability. Biogeochemistry 49:175–190. https://doi.org/10.1023/A:1006316117817

Ordentlich A, Elad Y, Chet I (1988) The role of chitinase of *Serratia marcescens* in biocontrol of *Sclerotium rolfsii*. Phytopathology 78:84–88

Pajares S, Gallardo JF, Masciandaro G et al (2011) Enzyme activity as an indicator of soil quality changes in degraded cultivated Acrisols in the Mexican Transvolcanic Belt. Land Degrad Dev 22(3):373–381. https://doi.org/10.1002/ldr.992

Paul EA (2007) Soil microbiology, ecology and biochemistry. Academic, Amsterdam

Peoples MS, Koide RT (2012) Considerations in the storage of soil samples for enzyme activity analysis. Appl Soil Ecol 62:98–102. https://doi.org/10.1016/j.apsoil.2012.08.002

Prietzel J (2001) Arylsulfatase activities in soils of the Black Forest/Germany: seasonal variation and effect of $(NH_4)_2SO_4$ fertilization. Soil Biol Biochem 33:1317–1328. https://doi.org/10.1016/S0038-0717(01)00037-2

Rao MA, Scelza R, Gianfreda L (2014) Soil enzymes. In: Gianfreda L, Rao MA (eds) Enzymes in agricultural sciences. OMICS Group, Foster City, pp 01–18

Sardans J, Penuelas J (2005) Drought decreases soil enzyme activity in a Mediterranean Quercus ilex L. forest. Soil Biol Biochem 37:455–461. https://doi.org/10.1016/j.soilbio.2004.08.004

Schneider T, Keiblinger KM, Schmid E et al (2012) Who is who in litter decomposition? Metaproteomics reveals major microbial players and their biogeochemical functions. ISME J 6:1749–1762. https://doi.org/10.1038/ismej.2012.11

Shapira R, Ordentlich A, Chet I et al (1989) Control of plant diseases by chitinase expressed from cloned DNA in *Escherichia coli*. Phytopathology 79:1246–1249

Sharma S, Khan FG, Qazi GN (2010) Molecular cloning and characterization of amylase from soil metagenomic library derived from Northwestern Himalayas. Appl Microbiol Biotechnol 86:1821–1828. https://doi.org/10.1007/s00253-009-2404-y

Sinsabaugh RL, Antibus RK, Linkins AE (1991) An enzymic approach to the analysis of microbial activity during plant litter decomposition. Agric Ecosyst Environ 34:43–54. https://doi.org/10.1016/0167-8809(91)90092-C

Sinsabaugh RL, Lauber CL, Weintraub MN et al (2008) Stoichiometry of soil enzyme activity at global scale. Ecol Lett 11:1252–1264. https://doi.org/10.1016/0167-8809(91)90092-C

Sinsabaugh R, Hill B, Shah J (2009) Ecoenzymatic stoichiometry of microbial organic nutrient acquisition in soil and sediment. Nature 462:795–799. https://doi.org/10.1038/nature08632

Šnajdr J, Valášková V, Merhautová T et al (2008a) Activity and spatial distribution of lignocellulose-degrading enzymes during forest soil colonization by saprotrophic basidiomycetes. Enzym Microb Technol 43:186–192. https://doi.org/10.1016/j.enzmictec.2007.11.008

Šnajdr J, Valášková V, Merhautová V et al (2008b) Spatial variability of enzyme activities and microbial biomass in the upper layers of *Quercus petraea* forest soil. Soil Biol Biochem 40:2068–2075. https://doi.org/10.1016/j.soilbio.2008.01.015

Šnajdr J, Dobiášová P, Větrovský T et al (2011) Saprotrophic basidiomycete mycelia and their interspecific interactions affect the spatial distribution of extracellular enzymes in soil. FEMS Microbiol Ecol 78:80–90. https://doi.org/10.1111/j.1574-6941.2011.01123.x

Spier TW, Ross DJ (1978) Soil phosphatases and sulphates. In: Burns RG (ed) Soil enzymes. Academic, New York, pp 197–250

Spohn M, Carminati A, Kuzyakov Y (2013) Soil zymography – a novel in situ method for mapping distribution of enzyme activity in soil. Soil Biol Biochem 58:275–280. https://doi.org/10.1016/j.soilbio.2012.12.004

Staben ML, Bezdicek DF, Smith JL et al (1997) Assessment of soil quality in conservation reserve program and wheat fallow soils. Soil Sci Soc Am J 61:124–130. https://doi.org/10.2136/sssaj1997.03615995006100010019x

Steinweg JM, Dukes JS, Wallenstein MD (2012) Modeling the effects of temperature and moisture on soil enzyme activity: linking laboratory assays to continuous field data. Soil Biol Biochem 55:85–92. https://doi.org/10.1016/j.soilbio.2012.06.015

Stone MM, Weiss MS, Goodale CL et al (2012) Temperature sensitivity of soil enzyme kinetics under N-fertilization in two temperate forests. Glob Chang Biol 18:1173–1184. https://doi.org/10.1111/j.1365-2486.2011.02545.x

Štursová M, Baldrian P (2011) Effects of soil properties and management on the activity of soil organic matter transforming enzymes and the quantification of soil-bound and free activity. Plant Soil 338:99–110. https://doi.org/10.1007/s11104-010-0296-3

Tabatabai MA (1994) Soil enzymes. In: Weaver RW, Angle JS, Bottomley PS (eds) Methods of soil analysis: microbiological and biochemical properties, Part 2, SSSA book Ser, vol 5. SSSA, Madison, pp 775–833

Trap J, Riah W, Akpa-Vinceslas M et al (2012) Improved effectiveness and efficiency in measuring soil enzymes as universal soil quality indicators using microplate fluorimetry. Soil Biol Biochem 45:98–101. https://doi.org/10.1016/j.soilbio.2011.10.010

Trasar-Cepeda C, Leiros MC, Gil-Sotres F (2000) Biochemical properties of acid soils under climax vegetation (Atlantic oakwood) in an area of the European temperate-humid zone (Galicia, NW Spain): specific parameters. Soil Biol Biochem 32:747–755. https://doi.org/10.1016/S0038-0717(99)00196-0

Valásková V, Baldrian P (2006) Degradation of cellulose and hemicelluloses by the brown rot fungus *Piptoporus betulinus* – production of extracellular enzymes and characterization of the major cellulases. Microbiology 152(Pt 12):3613–3622

Vancov T, Keen B (2009) Rapid isolation and high-throughput determination of cellulase and laminarinase activity in soils. J Microbiol Meth 79:174–177. https://doi.org/10.1016/j.mimet.2009.08.013

Venkatesan S, Senthurpandian VK (2006) Comparison of enzyme activity with depth under tea plantations and forested sites in South India. Geoderma 137:212–216. https://doi.org/10.1016/j.geoderma.2006.08.011

Vepsäläinen M (2001) Poor enzyme recovery by extraction from soils. Soil Biol Biochem 33:1131–1135. https://doi.org/10.1016/S0038-0717(00)00240-6

Verma D, Kawarabayasi Y, Miyazaki K et al (2013) Cloning, expression and characteristics of a novel alkalistable and thermostable xylanase encoding gene (Mxyl) retrieved from compost-soil metagenome. PLoS One 8:e52459. https://doi.org/10.1371/journal.pone.0052459

Wallenius K, Rita H, Simpanen S et al (2010) Sample storage for soil enzyme activity and bacterial community profiles. J Microbiol Methods 81:48–55. https://doi.org/10.1016/j.mimet.2010.01.021

Wallenstein MD, Mcmahon SK, Schimel JP (2009) Seasonal variation in enzyme activities and temperature sensitivities in Arctic tundra soils. Glob Chang Biol 15:1631–1639. https://doi.org/10.1111/j.1365-2486.2008.01819.x

Wang SL, Yen YH, Tsiao WJ et al (2002) Production of antimicrobial compounds by *Monascus purpureus* CCRC31499 using shrimp and crab shell powder as a carbon source. Enzym Microb Technol 31:337–344. https://doi.org/10.1016/S0141-0229(02)00135-7

Wang Y, Li Y, Ye X et al (2010) Profile storage of organic/inorganic carbon in soil: from forest to desert. Sci Total Environ 408:1925–1931. https://doi.org/10.1016/j.scitotenv.2010.01.015

Weand MP, Arthur MA, Lovett GM (2010) Effects of tree species and N additions on forest floor microbial communities and extra-cellular enzyme activities. Soil Biol Biochem 42:2161–2173. https://doi.org/10.1016/S0141-0229(02)00135-7

Wu QS, He XH, Zou YN et al (2012) Spatial distribution of glomalin-related soil protein and its relationships with root mycorrhization, soil aggregates, carbohydrates, activity of protease and β-glucosidase in the rhizosphere of Citrus unshiu. Soil Biol Biochem 45:181–183. https://doi.org/10.1016/j.soilbio.2011.10.002

Yuan B, Yue D (2012) Soil microbial and enzymatic activities across a chronosequence of chinese pine plantation development on the loess plateau of China. Pedosphere 22:1–12. https://doi.org/10.1016/S1002-0160(11)60186-0

Zahir ZA, Malik MAR, Arshad M (2001) Soil enzyme research. On Line J Biol Sci 1(5):299–307

Zavarzina AG, Leontievsky AA, Golovleva LA et al (2004) Biotransformation of soil humic acids by blue laccase of *Panus tigrinus* 8/18: an in vitro study. Soil Biol Biochem 36:359–369. https://doi.org/10.1016/j.soilbio.2003.10.010

Biosurfactants in Soil Bioremediation 11

Vivek Rangarajan and Mahesh Narayanan

Abstract

Biosurfactants are powerful surface active agents synthesized essentially by microbes. They have interesting properties such as biodegradability, less toxicity and stability at extremes of pH, temperature and salinity. Their diverse structures along with superior properties qualify them as potential candidates for application in food, cosmetic, pharmaceutical, agricultural and environmental industries. The current chapter discusses the salient features of two important biosurfactants, lipopeptides and rhamnolipids, and their use in the lab-scale remediation of soil contaminated with heavy metals and hydrocarbons.

Keywords

Biosurfactants · Lipopeptides · Rhamnolipids · Remediation of oil wastes · Heavy metal bioremediation

11.1 Introduction

The rise in world's population accompanied by the increase in pollution levels due to the discharge of wastes from industries and households has posed severe threat to the ecosystem, and this will greatly affect the mere survivability of the generations to come. The scientific communities around the world have taken a lot of effort towards mitigation of environmental pollution through renewable technologies. Although the classical,

V. Rangarajan (✉)
Department of Chemical Engineering, BITS Pilani, Sancoale, Goa, India
e-mail: vivekrangarajan@gmail.com

M. Narayanan
Department of Chemistry and Biosciences, Srinivasa Ramanujan Centre, SASTRA University, Kumbakonam, Tamil Nadu, India

T. K. Adhya et al. (eds.), *Advances in Soil Microbiology: Recent Trends and Future Prospects*, Microorganisms for Sustainability 3,
https://doi.org/10.1007/978-981-10-6178-3_11

physical and chemical routes of remediating the soil and waterbodies to reduce or remove the toxic pollutants have been very much effective, particularly, towards meeting the discharge standards set by various environmental regulatory boards, the cost involved in such chemical/physical treatments is high and often involves use of non-renewable and non-biodegradable chemical agents (Shah et al. 2016).

Of various pollutants contaminating soils, synthetic surfactants from petroleum industries hold a significant share. In general, they find applications as detergents, wetting agents, emulsifiers, foaming agents and dispersants in almost all fields spanning from agricultural to pharmaceutical industries (Rahman and Randhawa 2015; Somasundaran et al. 2004). Their increased consumption for various applications has resulted in increased discharge of them into waterbodies, thereby causing significant threat to the animals and humans. The use of synthetic surfactants as additives for bioremediation of contaminated soil with toxic pollutants such as heavy metals and hydrocarbons including polynuclear aromatic hydrocarbons (PAHs) (Ahn et al. 2009; Carroll and Campana 2008; da Silva et al. 2015; Dhenain et al. 2006; Ren et al. 2014; Souza et al. 2016) further adds up to the already existing demand of them towards the mainstream applications. Therefore, the use of eco-friendly and biodegradable green surfactants synthesized through microbial route is envisaged to reduce the footprint of the erstwhile used oil-based surfactants for such environmental applications, thereby making the entire process of remediation sustainable.

Biologically synthesized surfactants, also termed biosurfactants, are essentially synthesized by microorganisms. Based on their molecular weights, they are classified as low-molecular-weight and high-molecular-weight biosurfactants. Low-molecular-weight biosurfactants include glycolipids, lipopeptides and phospholipids, while high-molecular-weight biosurfactants include polymeric and particulate surfactants (Mukherjee 2007; Mukherjee et al. 2006). They are less toxic, highly biodegradable and stable at extremes of environmental conditions. Because of their superior properties such as high surface activity, emulsifying ability, anti-adhesive activity and antimicrobial and anticancer properties, they find applications in cosmetic industries, food processing, agricultural industries, health care and bioremediation (Banat et al. 2000; Gudina et al. 2013; Mukherjee et al. 2006; Mulligan 2005; Nitschke and Costa 2007; Rodrigues et al. 2006; Sen 2008). The powerful surface tension and interfacial tension reducing capabilities of green surfactants such as rhamnolipids and lipopeptides have already been exploited for application in enhanced oil recovery (EOR). Since biosurfactants are used for both low-end applications such as in EOR and bioremediation fields and for high-end applications such as in cosmetic and pharmaceutical industries, their production processes have to be strategized suitably in order to match the desired levels of purities. For example, biosurfactants used for environmental industries can be produced from low-cost substrates followed by simple purification steps to make the production process economically viable, whereas biosurfactants for pharmaceutical applications should be of high purity and therefore no compromise with the quality of raw materials and production process is affordable (Rangarajan and Clarke 2015). In the current chapter, the bioremediation potential of rhamnolipids and lipopeptides, two important classes of biosurfactants synthesized predominantly by

Pseudomonas spp. and *Bacillus* spp., respectively, is discussed. Firstly, the properties of these biosurfactants that qualify them as potential bioremediation candidate are discussed, followed by a brief discussion about lab-scale studies that demonstrate the successful use of biosurfactants in bioremediation of crude oil and heavy metals in soil.

11.2 Biosurfactants: Important Physicochemical Properties

11.2.1 Lipopeptides

Lipopeptides are amphiphilic molecules having a cyclic peptide as the hydrophilic group linked to a hydrocarbon chain, which acts as a hydrophobic moiety. Their diverse structures have resulted in numerous isoforms, which mainly belong to three important families, namely, surfactins, iturins and fengycins (Peypoux et al. 1999; Rangarajan and Clarke 2015). Most of the *Bacillus* spp. are capable of producing all the three types of lipopeptides but, at different concentrations, based on the medium and production conditions. *Bacillus subtilis* ATCC 21332 is the widely investigated strain for the production of surfactin, which is currently being marketed as a cosmetic agent by Kaneka Corporation, Japan, which can serve either as a bioactive compound if used in low quantity or as the principal emulsifying agent if used in excess in personal care products. As far as the structures of these compounds are concerned, surfactin has a cyclic lactone ring structure (Fig. 11.1) with C_{12}-C_{16} β-hydroxy fatty acid attached to a heptapeptide with a variable amino acid at positions 2, 4 and 7 (Bonmatin et al. 2003). Fengycin consists of a β-hydroxy fatty acid chain linked to a decapeptide. Fengycin A and fengycin B are the two variants with Val and Ala, respectively, at position 6 (Vanittanakom et al. 1986), and Iturin has a C_{14}-C_{17} β-amino fatty acid moiety linked to a cyclic heptapeptide moiety having Asp or Asn at position 1 (Bonmatin et al. 2003). While the structural diversity of these molecules leads them towards high-end therapeutic and cosmetic applications (Gudina et al. 2013; Mukherjee et al. 2006; Rangarajan and Clarke 2016), functional properties, such as surface activity and micelle-forming ability,

Fig. 11.1 Structure of surfactin

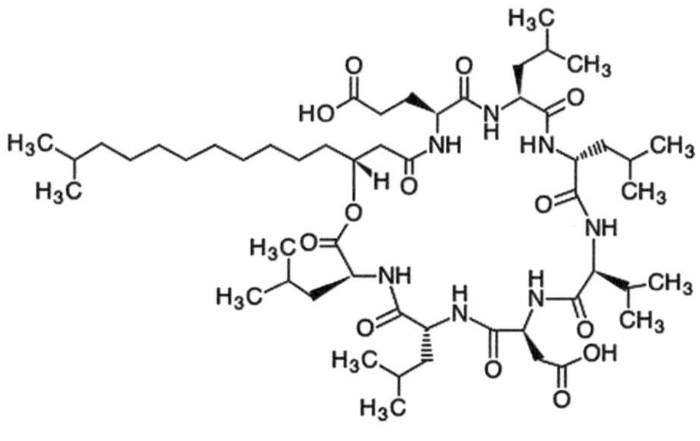

qualify these molecules for emulsion-related applications particularly in food and cosmetic industries and for application in environmental bioremediation (Das et al. 2009; Kanlayavattanakul and Lourith 2010; Mukherjee et al. 2006).

Of various important properties of lipopeptides, emulsifying abilities and heavy metal binding abilities are very vital for these surface active molecules to be considered for bioremediation applications. Of three lipopeptides investigated so far, surfactin, the widely tested one, has a critical micelle concentration (cmc) of 9.4 μM at pH 8.7 and at 25 °C with the minimum surface tension of 27.2 mN/m (Han et al. 2008; Sen and Swaminathan 1997; Shaligram and Singhal 2010). The structure of surfactin is so favourable that it can readily form oil-in-water microemulsions, with necessary conducive environmental conditions provided. These emulsion- and microemulsion-forming abilities together with detergent actions have already been exploited for EOR and hydrocarbon removal (Patel et al. 2015; Rangarajan and Clarke 2016). The two carboxylate groups of aspartic and glutamic acid residues act as bidentate for the binding of heavy metals towards them, thus making the molecule a strong sequestering agent for the removal of bivalent heavy metals such as Ni, Cd, Cu and Zn from the contaminated soil or water (Das et al. 2009; Mulligan et al. 1999b).

Lipopeptides are also being considered in the formulation of emulsions and microemulsions that cater to EOR, pharmaceutical and cosmetic applications. The important parameters such as critical packing parameter, CCP = 0.1435 and hydrophilic-lipophilic balance, HLB = 10–12 based on the chemical structure of surfactin lipopeptide indicate that these molecules are capable of forming spherical micelles and favour the formation of oil-in-water microemulsions, respectively (Gudina et al. 2013).

11.2.2 Rhamnolipids

Rhamnolipids are an important type of glycolipid biosurfactants that have been widely reported for oil spill bioremediation and removal of heavy metals from soil and water (De Almeida et al. 2016; Mukherjee et al. 2006; Randhawa and Rahman 2014; Wang and Mulligan 2004). The structure is composed of a β-hydroxy fatty acid linked to a rhamnose sugar molecule by the carboxyl end as shown in Fig. 11.2. They are

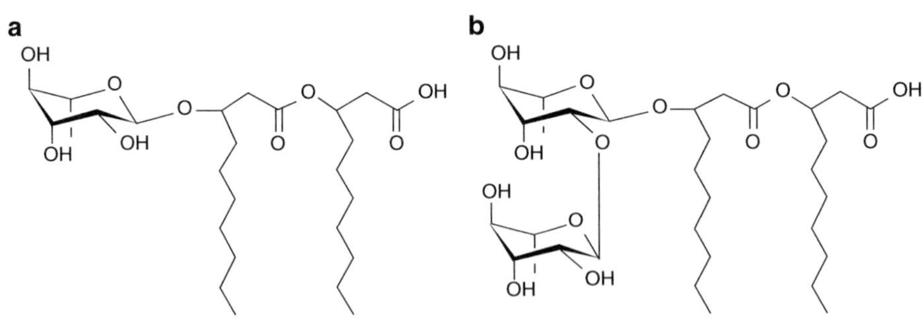

Fig. 11.2 Structure of (a) mono-rhanolipid and (b) di-rhamnolipid

predominantly produced by *Pseudomonas aeruginosa*. Based on the number of rhamnose sugars as hydrophilic groups, they are classified as mono and di-rhamnolipids (El Zeftawy and Mulligan 2011; Randhawa and Rahman 2014). Although rhamnolipid producers are capable of synthesizing both mono and di forms, the relative proportion of individual type can be varied by media optimization. As yield and productivity of rhamnolipids have been significantly improved over the last two decades through medium and process optimization, they have already made foray in many of applications such as in bioremediation and enhanced oil recovery, pharmaceutical industries, detergents and agriculture. The efficacy of rhamnolipids can be simply described by their ability to reduce the surface tension of water from 72.8 mN/m to about 29 mN/m (El Zeftawy and Mulligan 2011; Randhawa and Rahman 2014).

The only reason that rhamnolipid is still not being considered as the potential drug candidate for therapeutic application is due to the fact that most of the *Pseudomonas* spp. that are used to synthesize them are pathogenic to humans, plants and animals. The fact that *Pseudomonas* spp. are facultative anaerobes and they are capable of biodegrading PAHs and hydrocarbons, they can easily find applications in the wastewater treatment. However, again for the same reason that the pathogenic nature of these organisms prohibits them from being considered for such applications in larger scale.

11.3 Biosurfactants in Remediation of Oil-Contaminated Soil

The oil removal or recovery from the contaminated sand, works by three predominant mechanisms based on the concentration of biosurfactants used. The biosurfactant solution with a concentration < cmc helps in mobilizing the adhered oil, by the virtue of their surface tension and interfacial tension-reducing capabilities, while at high concentrations (>cmc) and very high concentrations, they are capable of mobilizing oil by solubilizing the oil or by forming emulsion with the oil phase. Here, the ability of biosurfactants to form emulsion with oil is determined by the molecular weight of biosurfactants, environmental parameters such as pH and temperature and the concentration and the amount of oil to be recovered from the contaminated site. The reclamation of oil-contaminated site can be discussed analogously to that of enhanced oil recovery by surfactants (Patel et al. 2015; Sen 2008), where surfactant solutions have been used as injection fluid in depleted oil wells (Fig. 11.3). Although the mechanism of action of biosurfactant in the recovery of oil from the contaminated site is similar to that of enhanced oil recovery procedure, the former one does not involve harsh environmental conditions such as elevated temperatures, which will influence the efficiency of the mobilization process. Also, the EOR involves pumping of injection fluid using high-power pumps to mobilize the fluid through interstitial spaces between the rocks, in order to mobilize the oil (Patel et al. 2015; Sen 2008).

The use of biosurfactants in EOR has been increasingly researched, from the view point of capturing more oil into the surfactant micelles. Micelles capture oil into the hydrophobic core in small quantities to form bigger micelles, also termed as swollen

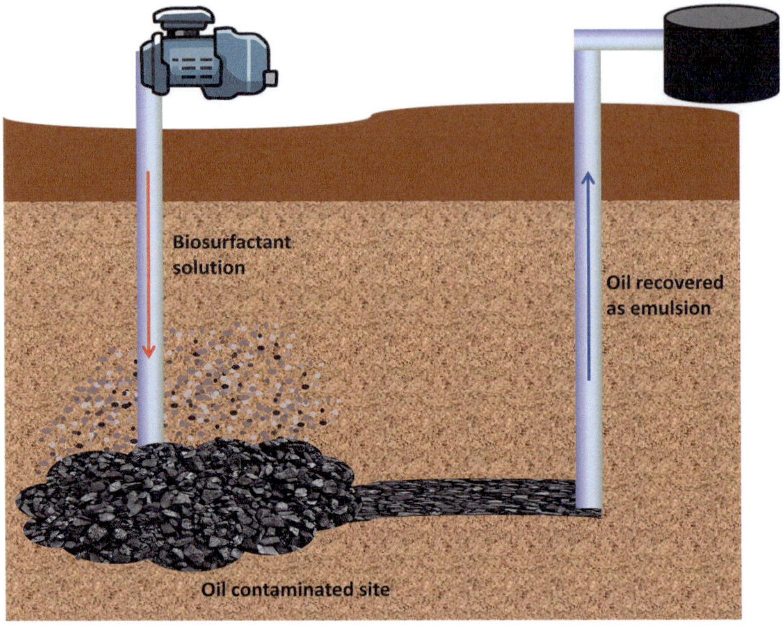

Fig. 11.3 Reclamation of oil-contaminated site by biosurfactant flushing (Adapted from Sen 2008)

micelles. The oil as discontinuous phase can also be finely dispersed inside the micelles, which usually maintains thermodynamic equilibrium with the continuous water phase termed as microemulsions. The classical EOR approach involves injecting biosurfactant as micelles/vesicles in flooding solution, which is devoid of oil, into the depleted oil well/oil- contaminated soil so as to capture oil into the micelles forming emulsions/microemulsions, while in recent approaches, biosurfactant prepared as microemulsions, i.e. micelles entrapped with some oil, is injected, which can help further entrapping more oil into them forming bigger swollen micelles, thereby recovering/remediating oil from the contaminated site (Bera and Mandal 2015). Microemulsion flooding is advantageous over classical surfactant flooding due to higher viscosity, ability to induce low surface tension and increased oil extraction efficiency (Santanna et al. 2009).

Formation of stable oil-in-water emulsion is understood to have improved effect on the microbial bioremediation of hydrocarbons. This improved bioavailability of oil/PAHs is usually achieved using biosurfactant-producing microbes, which facilitates microbes better access to oil. In a typical bio-stimulation approach, biosurfactant can be injected along with other required nutrients to enhance the oil biodegradation efficiency of microorganisms.

The investigation of optimal processing and environmental conditions is important for the effective oil removal. Urum et al. (2004) used Taguchi experimental design method and identified the optimal conditions as temperature, 50 C;

rhamnolipid concentration, 0.5% w/w; volume of solution, 15 mL; and contact time of 10 min for the effective removal of oil from the test contaminated soil.

Although biosurfactants are directly used to remove oil from the oil-contaminated sand, they can also be employed to remove the oil from the sorbents which are used to remove the oil from the contaminated site. In one reported study, Wei et al. (2005) investigated the efficiency of rhamnolipid biosurfactant for the removal of crude oil and weathered oil from the used polypropylene fibres as used sorbent. Sorbent pore size and washing time showed profound influence on the oil removal efficiency, whereas temperature and biosurfactant concentration showed little effect.

Chaprao et al. (2015) investigated the potential of two biosurfactants for their application in enhanced oil removal and biodegradation of motor oil-contaminated sand under laboratory conditions. The lipopeptides produced by *Bacillus* spp. removed 40% of the oil from the sand-packed column, while biosurfactant produced by *Candida* spp. removed as high as 90% of oil. Despite less recovery by lipopeptide, the degradation ability of *Bacillus* spp. was reported to be almost 100% after 90 days of incubation. In another work, Amani 2015 (Amani 2015) could achieve crude oil recoveries of 80 and 77% from the oil-contaminated sand at room temperature. Although most of the oil recovery has been carried out in simulated sand-packed columns, whose efficiencies, owing to the use of uniformly sized sand, have been reported to be high, a study by Colloney et al. (Connolly et al. 2010) showed poor/no recovery of oil from the real-time weathered oil-contaminated soil by rhamnolipid solution. The poor recovery was said to be attributed to the transfer of oil from the coarse sand particles to the fine clay component, which prevented the migration of oil from the sand to wash biosurfactant solution. Further studies with higher concentrations of biosurfactant are required to test the actual efficacy of biosurfactant to remove oil from weathered oil-contaminated sand.

11.4 Bioremediation of Heavy Metals Using Biosurfactants

Remediation of soil contaminated with heavy metals has been carried out using expensive conventional methods like excavation, land filling, isolation, immobilization, toxicity reduction, physical separation and extractions (Mulligan et al. 2001). Other promising alternative methods include soil washing, in situ flushing, bioleaching, phytoremediation and bioremediation. In recent years, the use of biosurfactants has been increasingly reported for the remediation of heavy metal-contaminated sand. As already stated before, anionic biosurfactants have superior structural and functional characteristics, which effectively remove the heavy metals from the adsorbed soil media through mechanisms such as electrostatic interaction, ion exchange, precipitation, dissolution and counterion binding. A reduction in interfacial tension facilitates binding of biosurfactants directly with the sorbed metals at interface. Anionic biosurfactants such as surfactin and rhamnolipid work effectively by forming complex with cationic heavy metal ions. The fact that interaction between biosurfactants and metals is stronger than the interaction

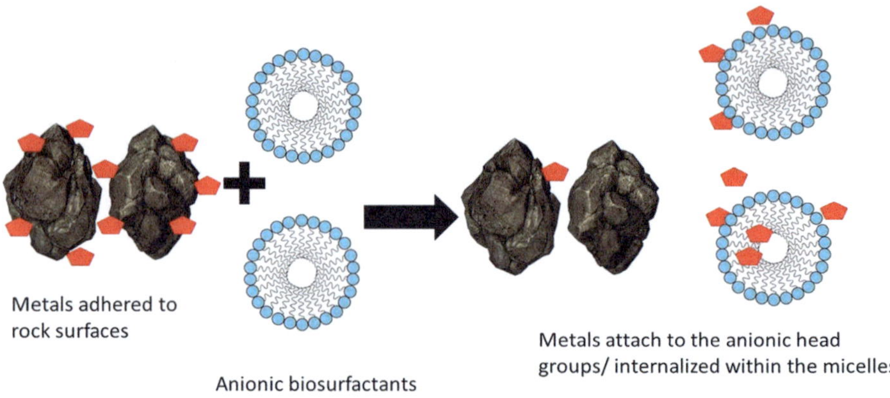

Fig. 11.4 Mechanism of metal removal by cationic biosurfactants

between heavy metals and sand facilitates easy disengagement of metals, which serves as the first line of mechanism in the removal of heavy metals from the sand (Fig. 11.4).

Mulligan et al. (1999a) investigated the feasibility of using surfactin, rhamnolipid and sophorolipid biosurfactants to remove heavy metals such as Zn and Cu from oil-contaminated sand. The use of 12% rhamnolipid solution recovered as much as 19.5% Zinc and over 25% Copper metal ions. The sequential extraction procedure using rhamnolipid and surfactin could effectively remove organically bound copper from the sand. Surfactin as low as 0.25% also removed 25% copper. It was reported that biosurfactants surfactin and rhamnolipids work effectively in alkaline pH conditions, while sophorolipids work effectively under acidic pH conditions in the presence of HCl, which could effectively remove oxides and carbonates of bound Zinc metal ions.

Usually, most of the reported studies for extracting of metals did not involve the use of direct or indirect mechanism of bioleaching. Diaz et al. (2015) used a synergistic approach, in which alternative cycles of treatment with rhamnolipids and bioleaching with mixed bacteria of *Acidithiobacillus thiooxidans* and *Acidithiobacillus ferrooxidans* were adopted. It was observed that bioleaching alone removed 50% of Zn and 19% Fe, while rhamnolipid treatment at low concentration of 0.4 g/L removed 11% Fe and 25% Zn and 19% Fe and 52% Zn at 1 g/L rhamnolipid. On the other hand, the combinatory treatment of bioleaching and biosurfactants reached up to 36% and 63 to 70% of Fe and Zn respectively. In another interesting study, the use of citric acid as additive in a mixed rhamnolipid-citric acid formulation improved the recovery of Pb and Cd along with the removal of the pesticide lindane (Wan et al. 2015).

Slizovskiy et al. (2011) used commercial rhamnolipid surfactant blend JBR-425 for the extraction of aged metals such as Zn, Cu, Pb and Cd from a soil subjected to more than 80 years of metal deposition. They were able to achieve concentration reduction of 39, 56, 68 and 43% for Zn, Cu, Pb and Cd, respectively. But progressive acidification with citric acid or ethylenediaminetetraacetic acid (EDTA) improved extraction efficiency to more than 95% for all four metals tested. The toxicity of heavy metals to earthworms was greatly reduced, such that the survival of earthworms was increased to 75% in the biosurfactant remediated soil, thus demonstrating less toxic nature of biosurfactant for bioremediation applications.

Mulligan et al. (1999c) in exploring the mechanism of metal removal by surfactin through technique of ultrafiltration and through the measurement of octanol-water partitioning and zeta potential indicated metal removal by sorption at the interface followed by complexation and incorporation into the biosurfactant micelles as the predominant mechanisms.

Most of the studies reported used biosurfactants solution as the injection liquid for the removal of heavy metals from soil. The use of biosurfactant foam can also be considered as a viable option for heavy metal removal, as the foam phase offers higher interfacial area than the liquid. However, the parameters such as foaminess and foam stability of the biosurfactant under the operating pH and temperature conditions and the pressure that the foam builds in the packed bed have to be assessed in order to design an effective foam-based metal removal process. With rhamnolipid biosurfactant foam obtained from a 0.5% rhamnolipid solution and with passage of about 20 pore volumes of solution into the sand-packed column, Mulligan and Wang (2006) could achieve removal of 73.2% Cd and 68.1% Ni, while the removal with rhamnolipid solution was 61.7% Cd and 51% Ni.

In another study (Bendaha et al. 2016), froth flotation technique involving rhamnolipids using sodium sulphide as activation agent was investigated for the remediation of heavy metal- contaminated soil. The sulfidization using Na_2S modified the particle surfaces to make them more exposed for the adsorption by biosurfactants.

The metal ions removed by the micelles can be concentrated using micellar-enhanced ultrafiltration (MEUF), where the membrane of suitable MWCO can be able to concentrate the micelles, while sending out the water through the membrane as permeate (Jung et al. 2008; Mungray et al. 2012). Vast literature discusses the procedure for conducting MEUF for the concurrent removal of heavy metals and concentration of the heavy metal solutions. The metal bound biosurfactant usually precipitates if the concentration of the biosurfactant exceeds certain critical value (Das et al. 2009). Usually, heavy metals tend to easily coagulate at acidic pH values than at basic pH values in case of biosurfactant solutions.

11.5 Conclusion

Bacterial biosurfactants hold greater potential for application in various fields owing to their important properties like biodegradability and less toxicity. Various lab-scale studies have demonstrated the remediating potential of biosurfactants for metals and oil in the contaminated soils. While metal removal exploits mainly the structural properties of the biosurfactants, in particular, the ability of biosurfactants to bind to the heavy metals and chelate with them, oil removal uses the emulsifying and surface tension-reducing abilities of biosurfactants. The studies have further indicated that biosurfactants are equally efficient as other synthetic surfactants for remediation purposes. Metal removal studies have been predominantly carried out using biosurfactant solution as the wash solution, while a few studies have also applied biosurfactant foam as the medium for the satisfactory removal of heavy metals. As far as the oil removal is concerned, biosurfactants were reported to be very much effective on non-weathered contaminated soil, while it is less effective in case of weathered-contaminated soil. Both rhamnolipids and lipopeptides have been tested to be very effective for both oil and heavy metal removal from the contaminated soil. While the demand for biosurfactants for applications in bioremediation and oil recovery application is increasing, the processing economy in producing these compounds still remains to be a constraint for the large-scale application of these green surfactants.

References

Ahn CK, Park D, Woo SH, Park JM (2009) Removal of cationic heavy metal from aqueous solution by activated carbon impregnated with anionic surfactants. J Hazard Mater 164(2–3):1130–1136. https://doi.org/10.1016/j.jhazmat.2008.09.036

Amani H (2015) Evaluation of biosurfactants and surfactants for crude oil contaminated sand washing. Pet Sci Technol 33(5):510–519. https://doi.org/10.1080/10916466.2014.999941

Banat IM, Makkar RS, Cameotra SS (2000) Potential commercial applications of microbial surfactants. Appl Microbiol Biotechnol 53(5):495–508

Bendaha MEA, Meddah B, Belaouni HA, Mokhtar M, Tirtouil A (2016) Removal of zinc and cadmium ions from contaminated soils with rhamnolipid biosurfactant produced by pseudomonas aeruginosa S7ps5. J Fund Appl Sci 8(3):1146–1165. https://doi.org/10.4314/jfas.v8i3.27

Bera A, Mandal A (2015) Microemulsions: a novel approach to enhanced oil recovery: a review. J Pet Explor Prod Technol 5(3):255–268. https://doi.org/10.1007/s13202-014-0139-5

Bonmatin JM, Laprevote O, Peypoux F (2003) Diversity among microbial cyclic lipopeptides: iturins and surfactins. Activity-structure relationships to design new bioactive agents. Comb Chem High T Scr 6(6):541–556

Carroll B, Campana N (2008) The removal of particles from oils by nonionic surfactants. J Dispers Sci Technol 29(10):1515–1519. https://doi.org/10.1080/01932690802316934

Chaprao MJ, Ferreira INS, Correa PF, Rufino RD, Luna JM, Silva EJ, Sarubbo LA (2015) Application of bacterial and yeast biosurfactants for enhanced removal and biodegradation of motor oil from contaminated sand. Electron J Biotechnol 18(6):471–479. https://doi.org/10.1016/j.ejbt.2015.09.005

Connolly HE, Rahman PKSM, Banat IM, Lord RA (2010) Resource recovery and reduction of oily hazardous wastes via biosurfactant washing and bioremediation. In: Płaza G (ed) Trends in bioremediation and phytoremediation. Research Signpost, Thiruvananthapuram

da Silva SS, Chiavone O, Neto ELD, Foletto EL (2015) Oil removal of oilfield-produced water by induced air flotation using nonionic surfactants. Desalin Water Treat 56(7):1802–1808. https://doi.org/10.1080/19443994.2014.958103

Das P, Mukherjee S, Sen R (2009) Biosurfactant of marine origin exhibiting heavy metal remediation properties. Bioresour Technol 100(20):4887–4890. https://doi.org/10.1016/j.biortech.2009.05.028

De Almeida DG, Da Silva RDFS, Luna JM, Rufino RD, Santos VA, Banat IM, Sarubbo LA (2016) Biosurfactants: promising molecules for petroleum biotechnology advances. Front Microbiol 7: Artn 1718. https://doi.org/10.3389/Fmicb.2016.01718

Dhenain A, Mercier G, Blais JF, Bergeron M (2006) PAH removal from black sludge from aluminium industry by flotation using non-ionic surfactants. Environ Technol 27 (9):1019–1030. https://doi.org/10.1080/09593332708618716

Diaz MA, De Ranson IU, Dorta B, Banat IM, Blazquez ML, Gonzalez F, Munoz JA, Ballester A (2015) Metal removal from contaminated soils through bioleaching with oxidizing bacteria and rhamnolipid biosurfactants. Soil Sediment Contam 24(1):16–29. https://doi.org/10.1080/15320383.2014.907239

El Zeftawy MAM, Mulligan CN (2011) Use of rhamnolipid to remove heavy metals from wastewater by micellar-enhanced ultrafiltration (MEUF). Sep Purif Technol 77(1):120–127. https://doi.org/10.1016/j.seppur.2010.11.030

Gudina EJ, Rangarajan V, Sen R, Rodrigues LR (2013) Potential therapeutic applications of biosurfactants. Trends Pharmacol Sci 34(12):667–675. https://doi.org/10.1016/j.tips.2013.10.002

Han YC, Huang X, Cao MW, Wang YL (2008) Micellization of surfactin and its effect on the aggregate conformation of Amyloid beta(1-40). J Phys Chem B 112(47):15195–15201. https://doi.org/10.1021/Jp805966x

Jung J, Yang JS, Kim SH, Yang JW (2008) Feasibility of micellar-enhanced ultrafiltration (MEUF) for the heavy metal removal in soil washing effluent. Desalination 222(1–3):202–211. https://doi.org/10.1016/j.desal.2007.01.154

Kanlayavattanakul M, Lourith N (2010) Lipopeptides in cosmetics. Int J Cosmet Sci 32(1):1–8. https://doi.org/10.1111/j.1468-2494.2009.00543.x

Mukherjee AK (2007) Potential application of cyclic lipopeptide biosurfactants produced by Bacillus subtilis strains in laundry detergent formulations. Lett Appl Microbiol 45 (3):330–335. https://doi.org/10.1111/j.1472-765X.2007.02197.x

Mukherjee S, Das P, Sen R (2006) Towards commercial production of microbial surfactants. Trends Biotechnol 24(11):509–515. https://doi.org/10.1016/j.tibtech.2006.09.005

Mulligan CN (2005) Environmental applications for biosurfactants. Environ Pollut 133 (2):183–198. https://doi.org/10.1016/j.envpol.2004.06.009

Mulligan CN, Wang SL (2006) Remediation of a heavy metal-contaminated soil by a rhamnolipid foam. Eng Geol 85(1–2):75–81. https://doi.org/10.1016/j.enggeo.2005.09.029

Mulligan CN, Yong RN, Gibbs BF (1999a) On the use of biosurfactants for the removal of heavy metals from oil-contaminated soil. Environ Prog 18(1):50–54. https://doi.org/10.1002/Ep.670180120

Mulligan CN, Yong RN, Gibbs BF (1999b) Removal of heavy metals from contaminated soil and sediments using the biosurfactant surfactin. J Soil Contam 8(2):231–254

Mulligan CN, Yong RN, Gibbs BF, James S, Bennett HPJ (1999c) Metal removal from contaminated soil and sediments by the biosurfactant surfactin. Environ Sci Technol 33 (21):3812–3820. https://doi.org/10.1021/Es9813055

Mulligan CN, Yong RN, Gibbs BF (2001) Surfactant-enhanced remediation of contaminated soil: a review. Eng Geol 60(1–4):371–380. https://doi.org/10.1016/S0013-7952(00)00117-4

Mungray AA, Kulkarni SV, Mungray AK (2012) Removal of heavy metals from wastewater using micellar enhanced ultrafiltration technique: a review. Cent Eur J Chem 10(1):27–46. https://doi.org/10.2478/s11532-011-0134-3

Nitschke M, Costa SGVAO (2007) Biosurfactants in food industry. Trends Food Sci Technol 18 (5):252–259. https://doi.org/10.1016/j.tifs.2007.01.002

Patel J, Borgohain S, Kumar M, Rangarajan V, Somasundaran P, Sen R (2015) Recent developments in microbial enhanced oil recovery. Renew Sust Energ Rev 52:1539–1558. https://doi.org/10.1016/j.rser.2015.07.135

Peypoux F, Bonmatin JM, Wallach J (1999) Recent trends in the biochemistry of surfactin. Appl Microbiol Biotechnol 51(5):553–563

Rahman PKSM, Randhawa KKS (2015) Editorial: microbiotechnology based surfactants and their applications. Front Microbiol 6:Artn 1344. https://doi.org/10.3389/Fmicb.2015.01344

Randhawa KKS, Rahman PKSM (2014) Rhamnolipid biosurfactants – past, present, and future scenario of global market. Front Microbiol 5:Artn 454. https://doi.org/10.3389/Fmicb.2014.00454

Rangarajan V, Clarke KG (2015) Process development and intensification for enhanced production of Bacillus lipopeptides. Biotechnol Genet Eng Rev 31(1–2):46–68. https://doi.org/10.1080/02648725.2016.1166335

Rangarajan V, Clarke KG (2016) Towards bacterial lipopeptide products for specific applications – a review of appropriate downstream processing schemes. Process Biochem 51(12):2176–2185. https://doi.org/10.1016/j.procbio.2016.08.026

Ren MM, Yuan XZ, Zhu Y, Huang HJ, Zeng GM, Li H, Chen M, Wang H, Chen CY, Lin NB (2014) Effect of different surfactants on removal efficiency of heavy metals in sewage sludge treated by a novel method combining bio-acidification with Fenton oxidation. J Cent South Univ 21(12):4623–4629. https://doi.org/10.1007/s11771-014-2469-3

Rodrigues L, Banat IM, Teixeira J, Oliveira R (2006) Biosurfactants: potential applications in medicine. J Antimicrob Chemother 57(4):609–618. https://doi.org/10.1093/Jac/Dkl024

Santanna VC, Curbelo FDS, Dantas TNC, Neto AAD, Albuquerque HS, Garnica AIC (2009) Microemulsion flooding for enhanced oil recovery. J Pet Sci Eng 66(3–4):117–120. https://doi.org/10.1016/j.petrol.2009.01.009

Sen R (2008) Biotechnology in petroleum recovery: the microbial EOR. Prog Energy Combust Sci 34(6):714–724. https://doi.org/10.1016/j.pecs.2008.05.001

Sen R, Swaminathan T (1997) Application of response-surface methodology to evaluate the optimum environmental conditions for the enhanced production of surfactin. Appl Microbiol Biotechnol 47(4):358–363

Shah A, Shahzad S, Munir A, Nadagouda MN, Khan GS, Shams DF, Dionysiou DD, Rana UA (2016) Micelles as soil and water decontamination agents. Chem Rev 116(10):6042–6074. https://doi.org/10.1021/acs.chemrev.6b00132

Shaligram NS, Singhal RS (2010) Surfactin – a review on biosynthesis, fermentation, purification and applications. Food Technol Biotechnol 48(2):119–134

Slizovskiy IB, Kelsey JW, Hatzinger PB (2011) Surfactant-facilitated remediation of metal-contaminated soils efficacy and toxicological consequences to earthworms. Environ Toxicol Chem 30(1):112–123. https://doi.org/10.1002/Etc.357

Somasundaran P, Chakraborty S, Qiang Q, Deo P, Wang J, Zhang R (2004) Surfactants, polymers and their nanoparticles for personal care applications. J Cosmet Sci 55:S1–S17

Souza RS, Porto PSS, Pintor AMA, Ruphuy G, Costa MF, Boaventura RAR, Vilar VJP (2016) New insights on the removal of mineral oil from oil-in-water emulsions using cork by-products: effect of salt and surfactants content. Chem Eng J 285:709–717. https://doi.org/10.1016/j.cej.2015.10.007

Urum K, Pekdemir T, Copur M (2004) Surfactants treatment of crude oil contaminated soils. J Colloid Interface Sci 276(2):456–464. https://doi.org/10.1016/j.jcis.2004.03.057

Vanittanakom N, Loeffler W, Koch U, Jung G (1986) Fengycin–a novel antifungal lipopeptide antibiotic produced by Bacillus subtilis F-29-3. J Antibiot 39(7):888–901

Wan JZ, Meng D, Long T, Ying RR, Ye M, Zhang ST, Li Q, Zhou Y, Lin YS (2015) Simultaneous removal of lindane, lead and cadmium from soils by rhamnolipids combined with citric acid. PLoS One 10(6):ARTN e0129978. https://doi.org/10.1371/journal.pone.0129978

Wang SL, Mulligan CN (2004) Rhamnolipid foam enhanced remediation of cadmium and nickel contaminated soil. Water Air Soil Pollut 157(1–4):315–330. https://doi.org/10.1023/B:Wate.0000038904.91977.F0

Wei QF, Mather RR, Fotheringham AF (2005) Oil removal from used sorbents using a biosurfactant. Bioresour Technol 96(3):331–334. https://doi.org/10.1016/j.biortech.2004.04.005

MIX
Papier aus verantwortungsvollen Quellen
Paper from responsible sources
FSC® C105338

If you have any concerns about our products,
you can contact us on
ProductSafety@springernature.com

In case Publisher is established outside the EU,
the EU authorized representative is:
Springer Nature Customer Service Center GmbH
Europaplatz 3, 69115 Heidelberg, Germany

Printed by Libri Plureos GmbH
in Hamburg, Germany